D0571678

ADVENTURES FROM THE TECHNOLOGY UNDERGROUND

adventures from the technology UNDERGROUND

Catapults, Pulsejets, Rail Guns, Flamethrowers, Tesla Coils, Air Cannons and the Garage Warriors Who Love Them

William Gurstelle

Clarkson Potter/Publishers
New York

Published in the United States by Clarkson Potter/Publishers, an imprint of the Crown Publishing Group, a division of Random House, Inc., New York.
www.crownpublishing.com
www.clarksonpotter.com

Library of Congress Cataloging-in-Publication Data

Gurstelle, William.
Adventures from the technology underground : catapults, pulsejets, rail guns, flamethrowers, Tesla coils, air cannons, and the garage warriors who love them / William Gurstelle.
p. cm.
Includes bibliographical references and index.
1. Machine design. I. Title.
TJ230.G975 2005
621.8'15—dc22

2005020412

ISBN-13: 978-1-4000-5082-6
ISBN-10: 1-4000-5082-0

Printed in the United States of America

Design by Jane Treuhaft
Illustrations by Max Werner

10 9 8 7 6 5 4 3 2 1

First Edition

> This book is dedicated to Alice Gurstelle. There's never been a minute I wasn't glad you were my mom.

contents

preface

> A physical experiment which makes a bang is always worth more than a quiet one. Therefore a man cannot strongly enough ask of Heaven: if it wants to let him discover something, may it be something that makes a bang. It will resound into eternity.
>
> —GEORG CHRISTOPH LICHTENBERG

This book takes its name from Fyodor Dostoevsky's book *Notes from the Underground*. Dostoevsky's Underground is a dark cellar from which a nameless Underground Man writes his journal—a place both physically and philosophically apart from society's mainstream. And that's the way he wants it, because out of the mainstream, he can exert his own individuality. This is the only place where he can demonstrate that he is a creative and unique human being with creative and unique ideas and, most of all, a distinctive self.

Writing of his normal employment, the Underground Man states, "I had a sickly dread of being ridiculous, and so had a slavish passion for the conventional in everything external." But Underground, he can open up his creative side; he can be as articulate and understandable and self-revealing as he desires.

Dostoevsky's Underground is a dark, serious, and humorless place, which makes it much different from the Technology Underground, the general term used throughout this book to describe the society of amateur inventors and scientific enthusiasts whose activities are profiled here. The Technology Underground is a buoyant place, full of enthusiasm, joy, and camaraderie. Still, the word *Underground* fits both places well, for several reasons.

The Technology Underground provides the same sort of opportunity for expression as the windowless cellar provides for the Underground Man of St. Petersburg. This is an important similarity—in the Underground, all are free to express themselves without reserve, whether expression occurs through words or actions, through journals or through orthogonal machine layouts and bolted steel constructions.

This is a place of alternative, clandestine, anti-establishment, and semisubversive culture. The possibilities and potentials inherent in this type of self-directed science and technology are the foundations for the creations of intelligent, innovative, and often slightly strange men and women who have their own notion of the best use of free time.

The people profiled here do what they do largely without regard to economics, profit, government regulation, corporate or academic practices, or even good taste. Most laws (except for those of physics), regulations, and building codes are skirted or ignored.

You can do that in the Technology Underground. Here, no one ever needs to justify his or her actions on the basis of normality or financial return.

And no one needs to fear being ridiculous.

introduction: the technology underground

This book is about the art and science of making catapults, pulse-jets, flamethrowers, Tesla coils, high-voltage electrostatic machines, high-power amateur rockets, air cannons, rail guns, fighting robots, magneformers, and more. It is as much about the people who make them as it is about the culture that engenders their creation. This book is the end result of a long and detailed exploration, a result of two years spent spelunking in the Technology Underground.

The knowledge for making devices such as these is not taught in any school, shared in instructor-led seminars, or obtained by reading mainstream science textbooks. Rather, this knowledge is acquired only by consorting closely and directly with ardent technophiles in their workshops and labs, by commingling with them at their competitions and gatherings, and by participating in obscure subject-specific Internet chat rooms.

Most of all, it is attained by using one's own eyes, hands, and brain to weld, machine, solder, bend, drill, plumb, design, break, tweak, experiment, and tinker. It requires spending more money than one would think reasonable on a hobby, and risking more of life and limb than a loved one would think prudent. In short, it requires a determined and long-term exploration of the Technology Underground.

WHAT IS THE TECHNOLOGY UNDERGROUND?

When is something a part of the Technology Underground? It is reasonable to outline, at the outset of a conceptualization or a project, the rules of conformance. Not every piece of interesting technological fabrication or eccentric scientist's folly belongs in the Technology Underground. There is a large number of possible candidates for inclusion; if the definition is stretched too far, the idea loses meaning—the boundaries become too dim and porous. So definitions and qualifications of a sort are required.

The changing nature and the continuous introduction of ever-newer technologies makes it a bit problematic to determine which projects belong and which do not. However, for the purposes of writing this book, the extreme and radical tinkerers of the Underground and their inventions all meet five conditions. These qualities are the membership criteria, the admittance ticket to the TU.

1. Projects Are Founded upon Physical Sciences

All experiments, inventions, or projects of the Underground are ultimately based on the work of physical scientists. Such projects use Newtonian physics and Faraday-style electricity and magnetism. This means that the projects here are built upon the science of men such as Archimedes, Galileo, Maxwell, Faraday, and Newton.* These projects are typically much closer to the hammer, wrench, welding rod, and soldering gun technologies of Thomas

---

* The ten most important scientists in history, according to writer Isaac Asimov, are:

SCIENTIST	CONTRIBUTION
1. Archimedes	machines and mechanics
2. Charles Darwin	theory of evolution
3. Albert Einstein	theory of relativity
4. Michael Faraday	electricity and magnetism
5. Galileo Galilei	gravity
6. Antoine Lavoisier	chemistry
7. James Clerk Maxwell	electricity and magnetism
8. Isaac Newton	laws of motion
9. Louis Pasteur	immunization and pasteurization
10. Ernest Rutherford	atomic theory

Of these ten, only Archimedes, Faraday, Galilei, Lavoisier, Maxwell, and Newton have direct applicability to the material in this book. Einstein's and Rutherford's contributions were to theoretical physics and Pasteur and Darwin's were to the biological sciences.

Edison, the Wright brothers, and Henry Ford than to the virtual, computational machines of Bill Gates and John von Neumann.

2. Amateur Standing

These are the projects of amateur scientists and technology enthusiasts. Qualifying projects must be motivated by love, not money. The projects in this realm are the work of individuals or at most small groups. There are no large corporate R&D teams, no government-sponsored laboratories, and there exists no real hope to profit financially or recoup monies spent. Here, technological prowess and profit have at best a distant relationship, and typically no relationship whatsoever.

3. Element of Danger

TU projects are always edgy and often dangerous. While not every project described in these pages is overtly dangerous, in point of fact, the majority do have some risky aspect or involvement. Almost all have some hazardous detail, some threatening characteristic, some sharp, hard, pointy thing moving at high velocity, an exposed hot surface, or a high-voltage shock risk. Always, there is some attribute that would cause your mother to prefer you trade participation in this activity for a safer one—say, steeplechasing or prizefighting.

4. High Energy

Such undertakings almost always involve the creation of large amounts of kinetic energy. Most examples here involve something moving at high speed. In fact, there is almost always a boom to start matters off, a whoosh, and finally a splat. In some cases, electrons or photons are used instead of more massive particles, but the high-kinetic-energy aspect is almost always present.

5. Recognition

These are social endeavors, not secret ones. The projects here are meant to be seen, criticized, and lauded. If recognition is not provided live and in person, then it exists through the Internet, secondhand reports, and word of mouth. The Underground existence is not a solitary one; indeed, it is a vibrant community that celebrates its members.

The attributes above provide a framework for determining when a work belongs and when it does not. Say, for instance, that two software developers design a video game involving death rays and flying cars. It has sexy graphics and lots of postmodern, Armageddon-like sound effects. Does such a project belong to the Technology Underground?

Probably not. While it may be worthy of recognition and perhaps culturally edgy, there is no solid connection to the science of Newton and Faraday. Nor is there any real element of danger. And the energy levels never exceed the puny wattage levels needed to spray electrons over a television screen.

Now consider a high-velocity, electromagnetic pulse weapon, developed by Lawrence Livermore Labs as part of a Defense Department initiative to research an anti-missile system.

It certainly qualifies from a kinetic-energy point of view, and its lineage from the work of Newton and the rest is clear. And it probably won't ever turn a profit unless Livermore branches out into international arms dealing. But it is not the work of amateurs, not by a long shot. And it is most likely a secret project, not a social one.

Finally, consider a group of friends who decide to build their own submarine. A working two-person submarine is a mechanical project of great complexity. It must be carefully designed in accordance with the laws of classical physics and fluid dynamics in order to make it swim beneath the water's surface. It is certainly dangerous, perhaps even foolhardy, depending on the fabrication skills of the inventors. It is expensive and will likely never produce even a dollar in revenue, much less any profit. Such a project meets the subjective criteria outlined here and therefore belongs in the Technology Underground.

WHY IS THERE NO FLYING CAR IN YOUR GARAGE?

As Governor, I shall seek investors who will bring their capital to Louisiana in an effort to design, develop, and eventually mass-produce an aeromobile. This Vehicle, which would revolutionize transportation in America, would be a cross between an ultra light aircraft and an automobile. The intended purpose is to create the ability of lift-off between 55 and 75 MPH, flying at low altitudes for short distances, and conceptually look similar to an Indy racecar.

—UNSUCCESSFUL 2003 LOUISIANA GUBERNATORIAL CANDIDATE PATRICK "LIVE WIRE" LANDRY*

If the qualifications for determining whether a person belongs in the Tech Underground seem too academic, there is an alternative—a simple litmus test that works fairly well.

Sometimes, all the people in the world can be divided into two camps—those with a particular worldview, and everybody else. There is a dichotomy at work—a world in black and white. Either you get it or you don't. Either you think that Jerry Lewis is funny or you don't. Either you think Apple computers are the only way to go or you don't. In any case, you cannot fathom the point of view of those who don't share yours.

Inhabitants of the Technology Underground and those of the technology mainstream diverge on the relative viability of the holy grail of amateur tinkerdom: the flying car. Undergrounders cannot see any compelling scientific or sociological reasons why flying cars can't be perfected. In fact, many of them can't figure out

* Patrick E. Landry first threw his hat in the political ring in 1999 in the Louisiana special election to fill the congressional seat of the Hon. Robert Livingston. Congressman Livingston left office after disclosure of marital infidelities ruined his opportunity to take over as Speaker of the U.S. House of Representatives. Landry, called "Live Wire" because of his background as an electrician, claimed that among his qualifications for high office was his virginity. That, his plan to nuke Baghdad, and his flying-car development platform got him many votes. In the 2003 governor's race, Landry finished eighth out of seventeen candidates.

why they don't have one already. Mainstream scientists, on the other hand, find the idea so fraught with problems that it is impossible to conceive on any level—engineering (too complex), safety (too dangerous), marketing (too expensive), and legal (an administrative impossibility).

The flying car idea didn't start with animated cartoons in the 1960s, although most baby boomers probably imagine such a thing as looking and acting like what George Jetson uses to drop off his daughter, Judy, at Orbit High. It's a concept that has been tossed around since airplanes were first invented.

Hollywood set-designer-turned-engineer Norman Bel Geddes came up with one of the first flying car concepts.* In 1927, Bel Geddes left his successful business in theatrical set design and turned his attention toward designing streamlined cars, ships, and locomotive engines. One of his most imaginative works came in 1929, when he sketched plans for an airborne ocean liner, one that included space for passengers to enjoy cruise-liner-like activities such as playing shuffleboard, dancing to a full orchestra, eating multicourse meals in elegant dining areas, and working out in a spacious, fully equipped gymnasium.

It never got off the ground.

On the other end of the size and complexity spectrum was Bel Geddes's idea for a flying car. His plans yielded a design for something that looked much like a 1940 Chevy coupe with wings welded onto the sides and the wheels replaced by a single rear-facing propeller.

It never got off the drawing board, either.

Since then, quite a few models have taken flight. One of the first and perhaps most successful was the ConvAIRCAR. This was not, strictly speaking, a product of the Technology Underground, since

---

* Besides being the father of the flying-car concept, Norman Bel Geddes was also the father of Barbara Bel Geddes, onetime Hollywood ingenue and, most notably, portrayer of Miss Ellie, matriarch of the Ewing clan on TV's *Dallas.*

its backers fervently hoped to produce it commercially and make scads of money. Just after World War II, the Consolidated Vultee Aircraft Company of San Diego, better known as Convair, decided that the time was right for a flying car, and work started on the ConvAIRCAR. Convair poured hundreds of thousands of dollars into developing a flying car prototype. On paper, the ConvAIRCAR was the marriage between an automobile and an airplane. It promised to revolutionize the daily commute for thousands, perhaps millions, of people.

"The market for this flying automobile will be far greater than a conventional light plane," Consolidated Vultee marketers said, "because the purchasers can obtain daily use from the car to get more out of his investment." The company estimated minimum sales of 160,000 units. Nearly every traveling salesman in the country would buy into the concept. The estimated price was around $1,500, and the attachments to affix wings were an additional cost.

But these attachments were what made the whole concept fly, literally and figuratively. After driving to the airport, a ConvAIRCAR owner got out and connected a flight unit (which included the pusher prop) in order to take off and fly. When the ConvAIRCAR landed at the next airport, the pilot/driver removed the detachable wings, stowed them somewhere, and drove on to the final destination.

In November 1947, a prototype ConvAIRCAR circled San Diego for about an hour and a half. It appeared for a brief time that the Consolidated Vultee aircraft company had actually produced the "fertile mule," that is, a hybrid with a viable future. But, in reality, a single successful test flight proves little. A few days after the first flight, a test pilot crash-landed the ConvAIRCAR on a dirt road (and walked away) when it ran out of gas. The only prototype of the ConvAIRCAR in existence was damaged beyond repair. And that's as far as that air car ever went, because the bad publicity and the high costs of manufacture closed the book on the project.

Next up was the Mizar, a true invention of the Technology Underground, but with a far more tragic case history. Henry Smolinski and Hal Blake founded Advanced Vehicle Engineers in Van Nuys, California, in the early 1970s. Their plan was straightforward and practical, and quite similar conceptually to the ConvAIRCAR. They took the top half, engine, and wings from a Cessna Skymaster single-motor aircraft and placed it in an attachable module that fit on rails set on top of a modified 1971 Ford Pinto. By melding the two disparate vehicles through a variety of attachment methods, they came up with a flying car—half Ford economy car and half high-wing airplane.

The Cessna's wings rested over the Pinto's roof and the airplane engine stacked up against the Pinto's hatchback. The device appears to have had two engines: the Ford engine that powered the car's wheels for ground travel and a Lycoming 540 aircraft engine that drove the prop during flight. Which one powered the vehicle depended on whether you were driving or flying.

Initially it worked fairly well (though the airplane engine failed on the maiden flight, and the pilot had to drive back to the hangar). Publicity for the Mizar was quite extensive; an advertising agency was hired, and flying car stories are generally pretty easy to get on the evening news.

> Planned as a dual-use vehicle to fly long-distance travel and then operate as a conventional automobile for local surface travel, here's how the Mitzar [*sic*] works. Equipped with its pusher-type aircraft engine, the Mitzar airframe will be kept on telescopic supports at a convenient airport. You drive the AVE-modified Pinto to the hangar and back the car under the airframe. A self-aligning track incorporated into both units makes attachment an easy job that requires less than two minutes to complete.
>
> All flight controls and instrument hookups will be made with an umbilical connection, while structural connections will be locked in place with self-locking high-strength pins in the structurally linked track assembly and wing support connections.
>
> —EXCERPT FROM *PETERSON'S COMPLETE FORD BOOK*, 3RD EDITION, 1973

The last line in that description is worth special attention, for in late 1973, Smolinski and Blake climbed aboard the Mizar prototype one last time and powered up the engines. No one knows what went on in the cockpit of the Mizar as it rolled down the runway during takeoff. But what is certain is that very shortly after it left the ground, the "self-locking high-strength pins" gave way and the flying car developers found themselves driving through the southern California sky in a suddenly wingless and decidedly non-airworthy Ford Pinto.

The tragic death of the two principal developers resulted in the end of the Mizar project. So the world still awaits the first practical flying car. Hearteningly to those who wait and want, there is always another one looming on the horizon, ready to take off from the Underground and attempt to fly into the big time.

If you understand the appeal of a flying car, you can probably understand the appeal of the Technology Underground. If you are hopeful that one day—not soon perhaps, but someday—there will be a flying car, then the motivation of the people in this book will likely be understandable. If you can't, then the Underground inhabitant's motivation and enthusiasm may seem trivial.

BATTLE SCARS AND WORSE

As the passage regarding the demise of the flying Pinto attests, people who play with fire sometimes get burned, and in the Technology Underground, this warning should be taken literally. Many of the attributes that make these projects interesting also make them hazardous. Even those who know what they're doing run risks.

Many robot builders, rocketeers, and machine art makers proudly show off some pretty nasty scars suffered when they got too close to the spinning blade or reciprocating hammer of a robot or the exhaust nozzle of a rocket. One mechanical artist lost four fingers on his hand and had some of the missing digits surgically replaced with toes. Worse, people have been killed or injured in explosions caused by the manufacture of chlorates and reactive metals, which are ingredients used in building high-power rocket engines.

Although most explorers of the Technology Underground take pains to behave responsibly, at times things go wrong, sometimes through bad luck, but more frequently through poor judgment.

> BBC NEWSWIRE, MONDAY, NOVEMBER 25, 2002
>
> A STUDENT DIED AFTER BEING FLUNG 100 FEET INTO THE AIR BY A HUGE CATAPULT. KOSTADINE ILIEV YANKOV, 19, WAS TAKING PART IN THE STUNT AT THE MIDDLEMOOR WATER PARK IN WOOLAVINGTON NEAR BRIDGWATER, SOMERSET, ON SUNDAY.
>
> HE SUSTAINED SERIOUS INJURIES AT THE SCENE AND WAS AIRLIFTED TO FRENCHAY HOSPITAL IN BRISTOL WHERE HE DIED AT 1930 GMT ON SUNDAY. THE 50-FEET-TALL CATAPULT IS DESIGNED TO FLING A PERSON ON TO A SAFETY NET WHICH IS SECURED ABOVE THE GROUND. BUT THE BULGARIAN-BORN STUDENT CRASHED TO THE GROUND BEFORE REACHING THE NET.

> *FORT WORTH STAR-TELEGRAM,* APRIL 16, 2003
>
> A DENTON TEEN-AGER WAS BLINDED THIS WEEKEND AND FACES RECONSTRUCTIVE SURGERY BECAUSE A TOY GUN HE WAS PLAYING WITH SHOT A FROG INTO HIS FACE.
>
> DANIEL BERRY, 17, WAS LOOKING DOWN THE BARREL OF A "POTATO GUN" [SO CALLED BECAUSE TYPICALLY IT IS USED TO SHOOT POTATOES] WHEN IT WENT OFF, HIS PARENTS, LISA AND CLIFFORD BERRY, SAID AT AN AFTERNOON NEWS CONFERENCE AT JOHN PETER SMITH HOSPITAL.

People can easily be electrocuted by high-voltage experiments gone wrong, clobbered by a misaimed hurling machine, or take a shrapnel hit from just about any ill-conceived project gone critical.

Said Sam Barros, the webmaster of Powerlabs.org, a well-known Technology Underground website: "This is something I don't discuss even with my close friends. It is an embarrassment to me; a rather spectacular failure where I managed not only to screw something up majorly, but also to get seriously injured in the process. But here it is, in the interest of showing people how you can 'know' what you are doing and then have something go so wrong you regret it for the rest of your life."

This book explores the making of all sorts of dodgy machines, from high-voltage, high-power electrical apparatus to vegetable-shooting plastic cannons. All of them can be dangerous. And here is a final word of advice for those who, after reading this book, may be considering making their own flamethrower: don't.

You've been warned. Now, let's go spelunking.

Author's Note: **How to Read This Book**

As you peruse these pages, you will notice that there are separated sections, delineated by borders and backgrounds, scattered throughout this book. The purpose of these sections (they begin with the subject heading, "The Technology of . . .") is to provide you with a higher level of technical detail about the particular devices and machines described in the chapter. While many readers may appreciate this more technical detail on the physics of history behind, say, catapults, high-powered rockets, or flamethrowers, other readers may prefer to forego so much arcana.

If formulas or mathematical descriptions give pause or, even worse, frighten you, feel free to skim or skip them completely. There is no test at the end; you will not be graded.

However, if you are the type of person who delights in the details, please enjoy. In fact, feel free to delve into the subject matter more deeply, by visiting the list of books I supply in "Further Reading" (page 220) as a starting point.

1. high-power ROCKETS

Your finger hovers over the red button, and you move the microphone close to your mouth. You test the public-address system and are relieved to find that it works: When you speak, your voice is clearly heard all over the firing range.

Several hundred feet away is the launch pad, and on it stands the culmination of many hundreds of hours of labor and many thousands of dollars of your hard-earned discretionary income. It is your rocket, a 15-foot-tall accurate scale model of an American early 1960s solid-fuel Pershing I nuclear ballistic missile. It is a machine that you designed and built from scratch.

Your rocket is loaded with two stages of powerful chemical engines. Like the original Pershing, your motive power comes from two stages of precisely packed chemical fuel arranged in solid form. Each rocket engine is designed such that after it ignites, the gas from the burning chemicals will issue rearward in a high-velocity, high-temperature stream from the ceramic nozzle and propel the rocket up toward the stratosphere. Your rocket will reach empyreal heights, tens of thousands of feet—if all goes well.

You pay rigid attention to the preflight checklist. So far, everything looks like a go. There are small indicator lamps on the firing controls that signal launch status, and the ignition lamp shows green. This means that you have a working circuit, and so when the Fire button is pushed, enough current will be sent through the thin metal wire rammed into the motor to heat it red hot and thereby initiate the self-sustaining chemical reaction that occurs within the main motor's combustion chamber.

The countdown begins. *Ten. Nine. Eight . . .* At zero, you push the button and instantly great plumes of white smoke surround the base of the rocket. For a moment, the rocket doesn't move, and you too hold your breath. Then suddenly it leaps toward the sky with neck-jerking acceleration. The noise from the launch comes a split second after you see it leave, and when the noise does come, it is nearly deafening. The rocket climbs 100, 200, 500, 1,000 feet, its speed escalating logarithmically as it ascends. It climbs and climbs, and it becomes difficult, then nearly impossible, and then totally impossible to see the rocket itself, although the smoke and nozzle fire remain visible.

Everyone congratulates you on a successful launch. There is applause and backslapping, high fives all around.

But the celebration is cut short by the sound of the range safety officer's warning horn: Whoop! Whoop! Whoop! The RSO's voice is plainly heard over the public-address system. "Attention! Look up! Look up! We have a rocket coming in hot!" This is not good for you. This is not good for anybody. In fact, this is trouble with a capital *T*.

What has happened is this: your rocket has two stages. The first stage consists of several large chemical rocket engines that lift the entire rocket for the initial or "booster" phase of the flight. When expended, the booster rocket falls away, and a second engine, mounted above it, is supposed to automatically ignite and continue powering the remaining components upward.

But the second stage, powered by its own very large engine, has ignited later than it was supposed to. In fact, it ignited after the rocket reached apogee and had already turned and begun to head back to earth. So the engine is not powering the rocket to fly up higher. Your rocket is being driven back down to earth not only by gravity, but also by the second-stage engine. There is a real danger that the rocket will reach the ground and your launch area before this engine is burned out and triggers the timed ejection charge that deploys the recovery parachutes.

The current situation is this: There is a very large and heavy rocket coming your way on an unpredictable descent path, and not just in free fall, but pushed by the thrust of a high-impulse, high-velocity, solid-fuel rocket engine.

>>>

This is LDRS, the country's—and probably the world's—largest annual gathering of high-power amateur rocket enthusiasts. From all over the world, eager rocketeers come for a long weekend's worth of home-brewed acceleration and conversations about rocketry.

LDRS is an acronym for Large Dangerous Rocket Ships. It's the place where the people who started out as boys and girls experimenting with Estes and Centuri model rockets go when they want to build really, *really* big rockets of their own.

LDRS is sponsored by a group called Tripoli, which is the largest organization of high-power rocket makers. There are scores of local chapters or "prefects" in locations across the world. This year, Tripoli has chosen the Panhandle of Texas Rocketry Prefect to host the big event. The local leadership has been busy for months turning a large patch of cow pasture into the nation's most active rocket launching area.

Rocketeers both need and love wide-open spaces—the wider the better. Amateur rocket builders, especially those who specialize in building the largest and most powerful rockets, want only a couple of things: a lot of flat, open, unpopulated land in which to recover their rockets after flight, and clear, sunny skies. This makes places such as Texas, Kansas, and the Canadian prairie provinces ideal spots for LDRS gatherings.

The launch site is south of Amarillo, straight down the Interstate to the tiny hamlet of Happy, Texas. At that point, the route to LDRS follows Texas Ranch Road 287 east, a long, straight, and uncrowded chunk of pavement that goes through territory so flat you can practically see the curvature of the earth.

At the end of the long drive is the LDRS launch site, a sprawling temporary compound of tents, launch pads, electronics, and people. The level, open venue is perfect for facilitating the retrieval of the hundreds of rockets that will eventually drift back to earth during the event, attached by elastic shock cords to large white parachutes. This particular site has the additional and highly valued quality of being well outside all commercial air lanes, so the airspace above it has no scheduled flights. Even so, the Tripoli organizers had to apply for a certificate of special clearance from the Federal Aviation Administration, allowing very-high-altitude rocket flights during the three days of the event.

Central Texas can be brutally hot and bright in July, and the tents and E-Z Ups set up by the rocketeers and vendors provide the only shade. This meet has the air of a large crafts fair, except that the vendor booths contain recovery chutes, rocket engine casings, altimeters, and launch towers instead of decorated ceramic pots and fabric wall hangings. The east side of the area is dominated by rows and rows of missile launching pads.

In this heat, people are not inclined to exert themselves if they can help it, so most simply wander around the dusty field, working on their projects, talking to one another, and pointing. Spectators at a large-scale high-power rocket launch do a lot of pointing—always toward the sky, arms extended at about 70 degrees to the horizon. Their fingers trace out the rocket's acceleration skyward and then fall back down to their sides as they watch it float down on the end of a parachute or two.

Temperature notwithstanding, for a few days the formerly

sleepy area becomes an energetic beehive of activity: smoke plumes and contrails constantly hanging like puffy ropes over the ranch, rockets roaring up, then silently floating down.

The great number of participants keeps several launching pads active. The pads with the biggest rockets are placed the farthest away from people, for it is not unusual for a rocket to blow up, or in rocket lingo, "CATO," on the pad, producing a shrapnel rainstorm.*

On the afternoon of the second day, a really big rocket, two and a half stories tall, stands erect on the far launch pad. It is a gracefully proportioned and aerodynamically shaped rocket and it is beautiful, at least to a high-power rocket enthusiast. Spectators and rocketeers alike press toward the safety fence to get into position for the best view.

This is the Athos II rocket, built by the Gates brothers of California. Athos II is a very large rocket with high-specific-impulse engines and will likely attain great heights. This launch is obviously going to involve significant velocity, complexity, and power. Athos's launch has been anticipated for quite some time, so the crowd near the safety fence is thick. People reach for their binoculars and position their cameras on tripods. Over the facility's loudspeaker, the launch control officer begins the countdown for one of the highlights of LDRS-21.

---

*In the world of high-power rocketry, a CATO is an event that involves an explosive and unusually spectacular motor failure. It is one where all the propellant is burned in a spontaneous and drastically time-compressed fashion. There are several varieties of CATO. The propellant can blow out the nozzle, which is a loud but basically benign occurrence. Or the explosion can occur in the vicinity of the rocket motor's upper end cap, which usually destroys the rocket's recovery parachute and instrumentation. The most dangerous CATO is called a casing rupture. It is a sudden breach in the sidewall of the rocket tube, and almost always destroys the rocket completely.

The etymology of the term *CATO* is uncertain. Many rocket enthusiasts say it is an acronym for "catastrophe at takeoff," "catastrophic abort on takeoff," or something similar.

THE TECHNOLOGY OF HIGH-POWER AMATEUR ROCKETRY

In the typical solid-fuel rocket, the rocket maker builds a fiberglass shell that houses the motor, the recovery system, and whatever sensors, cameras, or other payload is placed within.* But the bulk of the rocket's weight is contained in its powerful chemical engines. In and of themselves, rocket engines are marvelous things. Their most basic form goes back to first-millennium China, when crude black powder was stuffed into bamboo rockets and used to frighten the enemy's horses. A simple rocket engine is straightforward and easy to understand. There is chemical propellant packed inside a metal casing. The chemicals inside the motor burn, and as they do so, hot, expanding gas is produced. This gas rushes out the back of the motor through a nozzle and, as described in Isaac Newton's Third Law of Motion, the backward gush of the gas results in an equal and opposite forward thrust of the rocket body. Simple, yes. But hey, this *is* rocket science, and things get complicated quickly.

Small, commercially available model rocket motors consist of black-powder propellant pressed under tons of pressure into a hard, dense matrix called "grain." When the grain is ignited, the motor starts burning linearly, like a very fast-burning cigarette, from its back to its front. As it does so, it pushes hot gas out through a clay nozzle, and the rocket zips forward until the propellant is all burned up.

The world of high-power rocketry is different and much more complicated. Instead of using a simple black-powder chemical rocket motor, the experienced flyers most often use engines made out of "composite propellant"—a combination of an oxidizer chemical such as ammonium perchlorate (AP) and a synthetic rubber binder material to hold the oxidizer in a desired shape and pro-

*Tripoli regulations forbid the inclusion of mice, hamsters, frogs, or small children as payload.

vide fuel. In addition, the rocket engine maker may mix in plasticizers, catalysts, and crosslinkers, all of which can make the propellant burn stronger, longer, slower, or hotter, depending on the goals of the rocket designer. Composite motors are formed into various shapes with voids and holes precisely designed into the motor in order to shape the direction and velocity of the exiting gas. Such complex contours and figures are complicated to fabricate, requiring great quantities of heating, molding, curing, machining, and, above all, attention to detail.

The most extreme rocket makers spend days on end experimenting with rocket designs and motor formulations. There are so many variables that the maker can adjust to affect the performance of the rocket. A quick list of their concerns includes the shape of the rocket body, fin design, the shape of the nozzle, the geometry of the motor's core, the combination of various chemicals that make up the propellant mixture, the rate of burn, and the ignition method. It takes a lot of scientific, mechanical, and seemingly alchemical knowledge to become a really good rocket maker. There is also an element of danger working with toxic and flammable chemicals such as ammonium perchlorate, potassium nitrate, and liquid oxygen.*

---

*Former NASA engineer Homer Hickam wrote a popular book called *Rocket Boys,* and from it came the enjoyable movie anagrammatically titled *October Sky.* Both told the story of four teenage boys from Coalwood, West Virginia, who in the late 1950s designed and built a homemade rocket that flew nearly 6 miles into the sky. In one scene, Hickam and his friends start building their rocket engine by tamping down gunpowder encased in a metal pipe. This was *not* a good idea—and something far worse than property damage could have resulted. But the boys learned from their experience and eventually figured out the difference between building rocket engines and building explosive devices.

Apparently Hickam's experience was not peculiar to him and his friends. People, especially teenage boys, went rocket-crazy in the late 1950s, their interest spurred to stratospheric levels during the patriotic frenzy caused by the launch of the Soviet Sputnik satellite. In the late fifties and early sixties, thousands of young people attempted to build rockets. Few had any idea of what they were doing, so most wound up building what in reality were pipe bombs. Unarguably, mixing inexperience, a surplus of enthusiasm, and powerful chemicals resulted in a dangerous situation.

> > >

What rocket makers care about most is the physics quantity called "total impulse." Total impulse is the product of the force acting on a rocket (the thrust) multiplied by the amount of time the thrust is applied. Expressed mathematically, it is:

Total Impulse = Average Thrust × Burn Time

An engine that applies a lot of thrust, for a long period of time, is a high-performance engine. To a rocket engine maker, the goal is lots and lots of impulse.

The size of a rocket motor and the amount of total impulse it produces are described by assigning the motor a letter of the alphabet. The smallest rocket motor is an A and is commonly sold in hobby stores without need for a permit. The B motor is twice as big as an A, and a C is twice as big as a B. Each increase in letter size denotes a doubling of the engine's rocket-lifting ability, or total impulse. The total impulse of an A-motor is about 2.5 newton-seconds (N-s), which is enough to lift a small rocket a few hundred feet. A B-motor provides 5 N-s, C-motors provide 10 N-s, and so on. The largest commercially produced rocket motor available to

---

Estes Industries, the biggest name in manufactured model rocket engines, published a booklet called *The Rocketeer's Guide to Avoiding Suicide.* This booklet labeled those who engaged in the activity of building homemade rockets as "Basement Bombers." It urged people never to make their own engines but instead to buy them premade.

According to Estes, about two hundred of the thousand people who responded to their survey said that a homemade rocket engine had caused serious injury to either them or someone they knew. The booklet provides example after example of rocket engine explosion injuries, some presented in gruesome detail: "He was making rockets out of pipes filled with match heads. The pipe blew up and he almost blew his stomach and intestines out. . . . He lost half a year of school."

Perhaps Estes had a vested interest in persuading young rocketeers to buy their products instead of building engines from scratch, but irrespective of that, there did seem to be an extraordinarily high accident rate among youthful rocket builders of the time.

certified amateur flyers, the mammoth N motor, provides a muscular 41,000 N-s. Custom engines are available from a number of boutique rocket engine designers. Some of these go into the O and P range and even beyond. They are large and energetic enough to power a half-ton rocket to jet-fighter altitudes. (Using this scale, the NASA space shuttle's 8.3 million Newton-second booster rockets are about two letters beyond a Z-motor.)

Although there are many variations in the design and construction of homemade rocket engines, one of the clearest differentiating factors is the type of chemicals used to provide the energy and hence the impulse. The two most common general categories of chemicals are those involving variations of black powder and those that use composite propellant. Composite engines are, pound for pound, significantly more powerful than black-powder engines, that is, they have a higher specific impulse.

Every rocket engine, from black powder to solid fuel composite to liquid fuel to hybrid systems, works in similar fashion and is subject to the same basic physical laws: The propellant is ignited. It burns. Hot and expanding gases are produced and then stream out of a nozzle. Thrust is produced and the rocket and whatever is attached to it goes forward.*

The force produced by the gas issuing out of the nozzle is called "momentum thrust." Imagine that a rocket engine builder constructs an engine with a burn rate of 10 pounds of fuel per second. Now further assume that the builder's rocket engine handbook tells him that his choice of rocket fuels will result in the gases leaving the rocket nozzle at a velocity of around 3,000 feet per second.

---

*But why does the rocket tube go forward? Two reasons. The first is because of the immutable physical law called "conservation of momentum," described by Isaac Newton in his Laws of Motion. Imagine a man sitting in a small boat in a still pond with a lapful of baseballs. He starts throwing the baseballs toward the back of the boat. Every time he does so, the boat goes forward. The harder he throws, the farther forward the boat goes. If he throws one ball after another, the boat moves continuously in the cpposite direction.

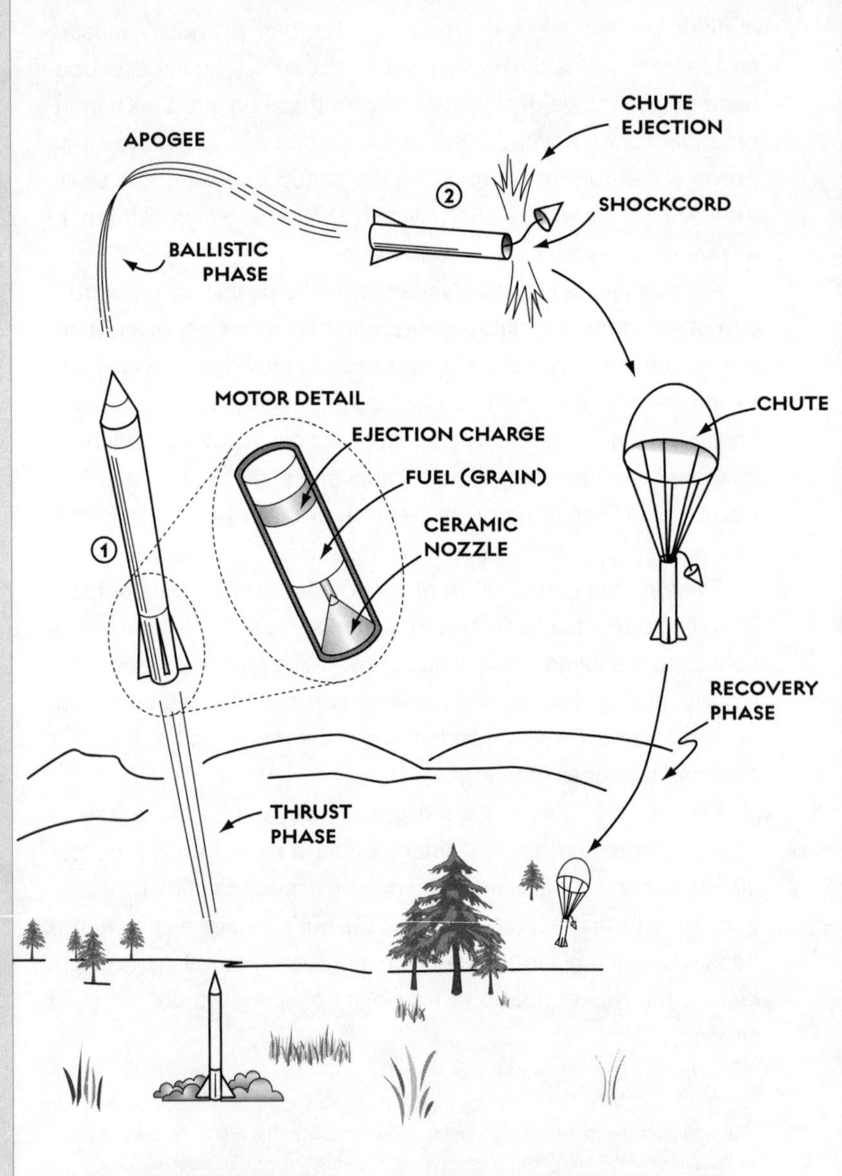

1. Motor burns bottom to top, through fuel. Burn continues without thrust through delay, then sets off ejection charge.

2. Ejection charge pushes wadding and chute against nosecone. Separation occurs and chute is ejected.

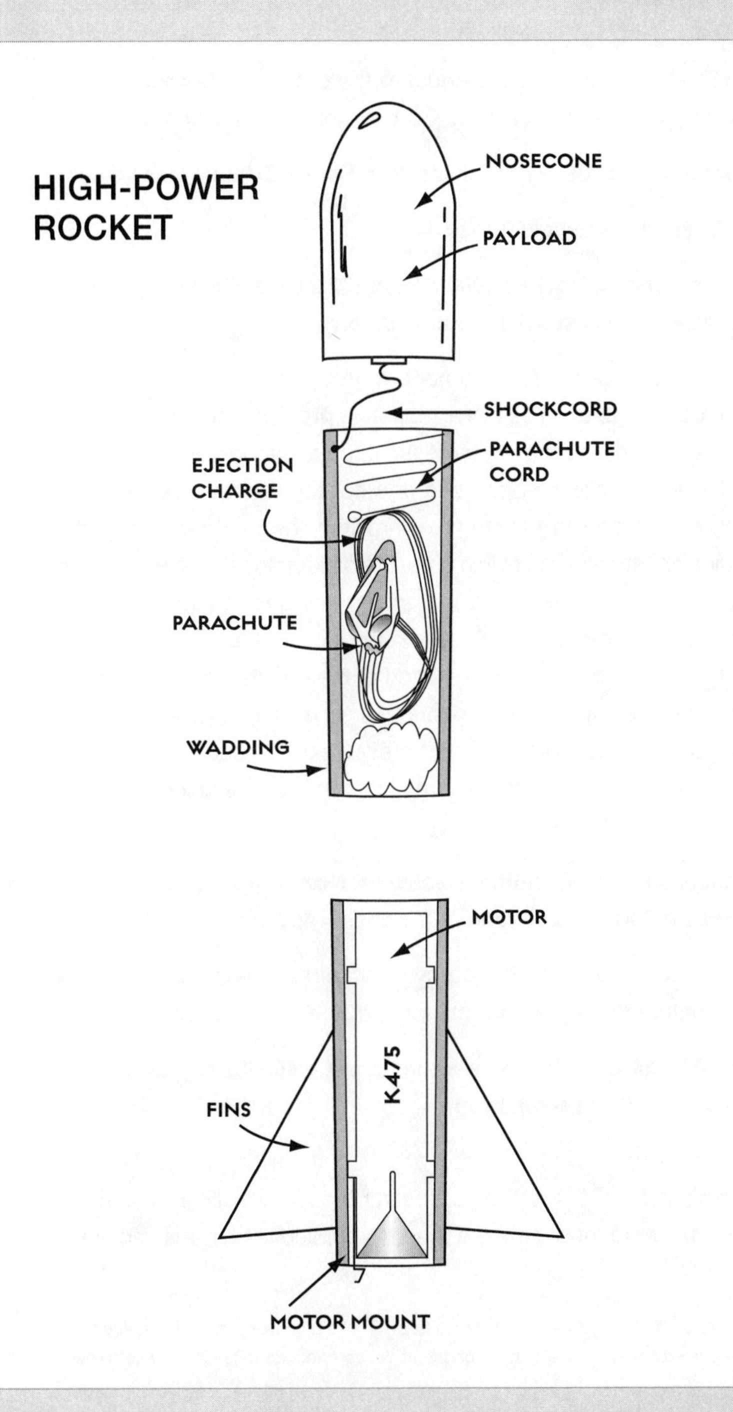
HIGH-POWER
ROCKET
NOSECONE
PAYLOAD
SHOCKCORD
PARACHUTE
CORD
EJECTION
CHARGE
PARACHUTE
WADDING
MOTOR
K475
FINS
MOTOR MOUNT

The thrust produced is equal to the propellant burn rate multiplied by the exhaust velocity. So the momentum thrust is:

Momentum Thrust = Propellant Burn Rate × Exhaust Gas Velocity

So, in the example above,

Thrust = (10 lbs/sec) × (3,000 ft/sec)/(32.2 ft/sec^2)*
Momentum Thrust = 932 pounds of force

So far, so good. But momentum thrust is only part of the reason rockets go up. The other reason is pressure thrust.

Inside a rocket engine, there are unbalanced forces at work. The rocket engine has an open end (the nozzle where the gases come out) and a closed end. During the burn time, the combustion of rocket engine chemicals results in a pressure buildup inside the engine. But since one end is closed and one end is open through the exit nozzle, there is a net force pushing against the closed end.

For example, assume the action of the burning chemicals inside the engine results in the production of expanding gas, which in turn results in a combustion chamber pressure of 200 pounds per square inch. If the exit nozzle has an area of, say, 2 square inches, then the pressure thrust is equal to:

Pressure Thrust = Engine Pressure × Nozzle Area
Pressure Thrust = 200 lb/in^2 × 2 inches = 400 lbs

The total thrust produced by a rocket engine is the sum of the momentum thrust and the pressure thrust. In this example,

Total Thrust = 932 lbs momentum thrust + 400 lbs pressure thrust = 1,332 lbs total thrust

Finally, consider the amount of time that the thrust is applied. The longer the time, the farther and faster the rocket will go. The thrust times the amount of time the thrust is applied is the total impulse.

---

*Where did that 32.2 ft/sec^2 term come from? That's a conversion constant due to the acceleration of gravity and is needed to convert pounds-mass to pounds-force.

I_{total} = **Thrust × Time**

So what does all this mean to the rocket designer? Plenty. The higher the propellant flow rate, the greater the thrust. The higher the velocity of the exhaust gases, the greater the thrust. The higher the combustion chamber pressure inside the engine, the higher the thrust. Taken together, there's a myriad of ways to increase impulse for any rocket.

But none of these factors is independent from the other factors. For example, the higher the flow rate, the shorter the duration of the thrust. A larger exit nozzle opening will result in more nozzle area but simultaneously less velocity in the exhaust gas stream.

Putting all of these things together is a challenging intellectual exercise, full of variables, trade-offs, and optimizations. And that in a nutshell is why rocket science is so darn complicated. <

THE ROCKETMEN

One of the best-known and most extreme rocket engine makers working in the Underground today is Frank Kosdon. Outwardly, he projects a nonconformist image and appearance: He is a tall man of indeterminate age with a napiform torso and thin legs, and his sartorial style tends toward ragged cutoff shorts and stained, shrunken T-shirts. His gray-black hair is beyond disheveled. Frank often fumbles his speech, and sometimes he struggles for words, but his eyes are bright and intelligent.

Frank earned his undergraduate degree in physics from Princeton and has a University of California doctorate. But he has chosen a lifestyle and career path distinctly different from those of most of his academic peers.

Kosdon has held various jobs. At the time of the Amarillo LDRS event, he earned most of his income in a couple of ways, both of which cause him to be on less than favorable terms with the California authorities. To meet "operational expenses" he often sells soda pop, beer, and wine coolers from the trunk of his Ford at the local nude beach. The local police don't condone either the nudity or the wildcat soda stand, and so there is friction.

The other thing Kosdon does for money is compound high-power rocket engines in his garage. Kosdon's motors are powerful and dependable, and provide a lot of bang for the buck. He has worked out all the intricate details—the nozzle geometry, the chemical composition, the casing design, everything required to make a powerful, hardworking chemical solid-fuel rocket motor. More than a few builders feel that Kosdon's solid-fuel motors are the Cadillacs of the high-power amateur rocket world.

At the time of the Amarillo launch, Kosdon's motors were high-demand items. In the early 1990s, a Kosdon-built rocket, launched from the Black Rock Desert near Reno, Nevada, held the non-government-entity world altitude record with an undocumented but widely believed height of 34,000 feet, which stood until fairly recently. But no Kosdon motors were flown at LDRS 21. There were some, well, issues.

Kosdon motors in the LMNOP range are said to be among the best, maybe the best. His fine craftsmanship aside, there is one big problem with Frank's motors, and it is this: His factory is located in his garage, which in turn is located in a densely populated Los Angeles suburb. Manufacturing rocket engines requires large quantities of chemicals—things such as the aforementioned ammonium perchlorate, black powder, metals such as aluminum and magnesium, plasticizers, epoxies, and other highly combustible items. Making a rocket motor requires heating, casting, and machining the chemicals. While this should not be overly dangerous when done by an expert such as Dr. Kosdon, such operations always present the possibility of an accident. A couple of years ago, a commercial high-power-rocket-motor factory blew up in a Las Vegas suburb, injuring several people and necessitating the evacuation of parts of the city.

Bowing to pressure from the Federal Bureau of Alcohol, Tobacco and Firearms, and the National Fire Protection Association, the organizers and safety committee of the LDRS launch won't allow Frank to sell his high-performance but non-ATF-certified motors at sanctioned meets such as this one.

At the launch site, one person asked Kosdon why he does business in the unauthorized and unorganized fashion he does. After all, a big rocket motor can cost more than $500, so this could be a viable business. But this is where rocket making starts to get a little political. A small but significant number of Underground members do not trust the government. Many of these are just moderately untrusting, but some of the more extreme don't trust it like, say, Idaho survivalists or Branch Davidians don't trust it. After a few beers, eventually some will start to explain. They talk of a broad national conspiracy on the part of ATF, NASA, the FAA, and others to stifle the activities of amateur rocket makers.

Why would the federal government give a hoot? It is because a few of the rocket men—among them the most extreme and the most talented—want to build *extremely* large rockets. And, according to some, building such large rockets is perceived as a threat to federal government interests. They believe the government will do everything it can to maintain its monopoly on space travel and commerce. The future is in space, they say, and the feds want it all for themselves.

"We're regulated and opposed at every turn," explained one of them. "The government put ridiculous and onerous rules and regulation out on everything from the purchase and use of chemicals like AP to making it impossible to fly our rockets at even relatively low altitudes like 20,000 feet, and even in the middle of the Nevada desert. Can someone explain to me why I can't go to, say, Australia or the middle of the ocean to launch my rockets? The government says that because we're U.S. citizens, they still have jurisdiction over our actions, even if we launch outside of Perth or from a boat beyond the 50-mile territorial limit. That's not fair, and it's not American. NASA wants to maintain their monopoly, and I think they'll stop at nothing to do so."

Even the much more numerous mainstream rocket flyers—those who don't have their own plans for exploiting outer space—often have issues with the federal government. The ATF took strong legal action in 2001 that amounted to a crackdown on the sport, enacting stringent rules regarding the storage, transport, and sale of the stuff that makes the whole activity go, ammonium perchlorate. Angered by the actions, rocket hobby associations such as Tripoli and the National Association of Rocketry started a legal battle with the Bureau over its regulations, and the battle continues. As usual, the only winners have been the lawyers.

So Frank Kosdon attends the Amarillo LDRS event strictly as an observer, since the event organizers don't want the government scrutiny that the use of his motors would cause. Instead of selling motors, he reverts to his other business interests, covering the cost of this trip to Texas by selling soda and beer from the trunk of his Ford.

Kosdon's home and workshop is located in a modest but well-tended section of the endless urbanized area that stretches north of Los Angeles toward Santa Barbara. Unlike his neighbors' houses, Frank's is not particularly well-tended or even neat. There are no window shades to hide the boxes upon boxes of rocket parts stored hither-skither in his living room. His house is so full of stuff that in most rooms there are only narrow aisles cleared for movement. Almost every cubic inch of space is filled with packing boxes, ceramic nozzles, the metal turnings that make up rocket motor cases, and above all, junk. Magazines, newspapers, and old telephone books, many dating from the early 1990s, form mazes of unstable columns reaching almost to the ceiling.

"I build rocket engines because I like it and because I'm good at it, mainly," he says. "It's hard to make money with this. The government does everything they can to stop a guy from making a living. But if you do still manage to succeed, they'll tax the hell out of you. Feel free to quote me on that."

Just a few miles down the Pacific coast from Frank Kosdon's wild jumble of a house is the home of another extreme tinkerer, an amateur rocket builder named Dirk Gates.

Dirk and Erik Gates are southern California brothers who are among the very highest flyers at LDRS. They build and fly some of the largest amateur rockets, including the Musketeers of the sky, Athos, Porthos, and Aramis. The Gates brothers possess the two things that are necessary to excel in the Technology Underground:

dedication and money. The Gateses, especially Dirk, have plenty of both because Dirk made a fortune in the overheated high-tech stock market of the late 1990s and sold his PC-card-making business, called Xircom, to Intel at just the right time.

Money, experience, and free time allow the Gates brothers to pursue their interest with a professional's competence and a hobbyist's ardor. Over time they have compiled an unparalleled panoply of large, high-power rockets that they truck around the country in a handsomely painted trailer. Professional stock-car racers and country western headliners should have a vehicle this nice.

Although the distance between Frank Kosdon's and Dirk Gates's homes is only 50 miles or so, the gap between them is vast. The approach to the Gates house is fronted by a massive, ornate steel gate, protected day and night by uniformed security men in a guardhouse. After the security checkpoint, a winding road flows past mansions reminiscent of fairy-tale castles. Gates lives in a California-style house that rambles on and on up the side of a hill. In a work space that spans several bays of the eight-stall garage, Dirk and his brother have set up the equipment to build some of the highest-flying rockets anywhere.

The Gates brothers make an unusual pair. Dirk is the quiet, unprepossessing one, but, no doubt about it, he's a driven engineer. A scorekeeper and technocrat, he built his computer company into a billion-dollar enterprise in a very short time. He works on his rockets and related paraphernalia in the garage, a 2,000-square-foot structure with 14-foot ceilings and composite-resin flooring instead of concrete, ringed neatly with cavernous white melamine cabinets holding tools, supplies, and miscellaneous sundries.

Erik Gates has different priorities. He appears more of a thrill seeker than a guy who loves business. Erik is a member of an elite bunch of just-outside-the-law adventurers who call themselves

"BASE jumpers" and are known for parachuting off things that are not airplanes—for instance, bridges or tall buildings. It's an underground fraternity, since most municipalities do not allow people to jump off office towers. Despite the frequently illegal nature of the activity, it is often practiced, and for people like Erik Gates it's a passion. *BASE* is actually an acronym for building, antenna, span (meaning a bridge), and earth (usually a tall cliff), the four different types of non-airplane parachute jumps that must be accomplished in order to be considered a true BASE jumper.

"Both Erik and I have always been interested in this kind of stuff," Dirk said. "Our dad was an aeronautical engineer, and we always tinkered around with him. But you can only go so far with your dad. In seventh grade, I found a book in my junior high library, believe it or not, that told me how to mix up a batch of gunpowder. I still remember how I felt when I found out that the secret to this stuff was to mix just the right proportion of three ingredients by weight together in a closed container, and then ignite it. We found sources for charcoal, sulfur, and saltpeter and ground them all down to a fine powder. We had a chemistry set at the time that had a burner that could melt and shape glass. So we put the gunpowder mixture inside a glass tube along with a Nichrome fuse and sealed the ends shut by softening and sealing the glass with the Bunsen burner."

Erik continued, "One of the neighbor kids had a dollhouse that was made from concrete blocks. We set the filled glass tube in there and set it off. There was a pretty big bang and we kind of wrecked the dollhouse, what with all the glass fragments and shrapnel.

"No police showed up. I wasn't surprised that they didn't, because it wasn't that big a deal. At least it wasn't back then, because people didn't seem to care so much about activities like this. The only time we got in real trouble was when we filled up a trash can with flammable gas. We used one of those *MacGyver*-type reactions to fill up a Hefty bag with hydrogen. Something to do with aluminum foil and household cleaners."

Dirk went on, "That was pretty clever for kids our age. When we lit it, we got a bigger reaction than we expected. It actually started a grass fire in our backyard and the fire department came out. We didn't get in trouble, though."

At the Amarillo LDRS launch site, the Gateses' Athos II roars into the sky with the force equivalent of a locomotive engine pushing it upward, with the fast-burning combination of ammonium perchlorate and plasticized fuel combusting and spilling madly out of the ring of exit nozzles. Tightly packed within the rocket engine array, each M-sized motor fired with approximately 1,400 to 1,500 pounds of force, putting the combined thrust in the neighborhood of 10,000 pounds. This is an amazing amount of thrust for a couple of amateur hobbyists to produce. During the period of main engine ignition, this lightweight rocket is pushing as hard as a small jetliner or a medium-sized tugboat. The orgy of thrust, flames, power, and smoke sends the rocket several thousand feet in just a few seconds.

When the booster rockets burn out, they fall away, and the main rocket body continues to climb in unpowered ballistic flight until the massive second-stage engines kick in. The second-stage motor, a combination of M- and J-sized engines, air-starts and sends the rocket farther still until finally gravity catches up and grabs it somewhere between four and five miles up. Then the rocket turns and begins its descent. Inside the rocket body, barometers sense the change in the rocket's direction and fire the parachute ejection charge, and the Athos II floats back down, exhausted but triumphant.

> > >

There's plenty of excitement at LDRS right now. Your big multistage rocket with the failed chute-ejection charge continues to barrel on down. It's descending fast and getting faster.

Life has its exposure risks, such as secondhand smoke, mosquito-borne viruses, and asbestos and radon lurking in your basement. Those who choose to attend high-power rocket meets must accept this additional one as well. The horns keep up their *whoop-whoop-whoop* as everyone nervously stares upward at the plume of smoke denoting the rocket plunging earthward on a kamikaze trajectory.

The range safety officer is the person responsible for the safe operation of the event, and he is very concerned. The rocket is large enough and heavy enough to wreck a car if it plows into one, and if it hits a person, that would be far worse. It's too late to evacuate, so the people on the ground must be ready to run.

Finally, and to everyone's great relief, the altimeters and angle sensors on board figure out what's going on. A puff of smoke appears in the sky, immediately followed by the preliminary or drogue chute, and then the main chute—a big white and orange pillow of cloth.

The rocket body slows rapidly and then floats down, pushed by gentle winds out to the cow pasture, where a group of young men on all-terrain vehicles retrieve the fuselage. Your rocket lives to fly again.

> > >

In 2004, the first homemade rocket to break the government monopoly on reaching the official limits of outer space (100 kilometers up) made its successful launch from a place in the northwestern corner of Nevada. Most of the time, there's not much going on there. It's an area known as the Black Rock Desert, a flat, dusty, windy, unpopulated dry lake bed, called the playa. Black Rock is a much-favored location for high-power amateur rocketeers to try out their largest and most powerful motors.

In 1997 the playa accommodated a supersonic land speed

attempt by the British SSC Thrust jet car, which hit a speed of 763 mph, or Mach 1.02.

Aside from technology record attempts, not much would seem to occur on the Black Rock Desert's playa to make it qualify as a high-energy hot spot—except for one particular time of the year. Then, for a week in late summer, the playa comes alive with unbelievable energy and activity.

2. the technology of BURNING MAN

Your experience started a week earlier, about four hundred miles west of the playa in the affluent suburbs surrounding San Francisco Bay. It started as an easy drive up I-80 through Sacramento and on past Reno. At the Wadsworth, Nevada, exit, you left the freeway and made the turn due north.

"Okay," says your friend in the co-pilot seat of your borrowed Volkswagen Vanagon, "here's the area where you really need to watch your speedometer closely."

The police are on high alert and strictly enforcing the traffic laws in the normally lax enforcement areas in Nevada's barren northwest corner. You've already seen quite a few cops along the roadside, but you've maintained legal speed and there have been no problems.

You are moving quite well, cruising along with the air-conditioning turned up high. Although it is getting toward sunset, it is still hot outside, and very windy to boot. You've been driving continuously for the last eight hours, going as fast as you can without getting a ticket, and staying on an east-northeast heading.

For the last several hours it's simply been a flat-out burn through the desert. To this point you've enjoyed taking in the scenery from the comfort of your air-conditioned VW. But after the next hour or so, that's the end of it, for the next seven days will be strictly out of doors. When the sun shines, it will be hotter than hot and there will be no AC to cool you off. Your friend has been here before, but this is your first time. You feel a little nervous about the whole thing. Was it a good idea to come here after all?

As you highball over twisty but smooth two-lane roads at freeway speeds, this singular desert shows itself as one of the most unusual landscapes in the United States. According to the local Bureau of Land Management, the Black Rock playa was formed from "a unique assemblage of volcanic lava flows, ash, ancient shallow marine sea floor, exotic batholithic terrain, and lacustral sedimentary packages." Translated into ordinary English, this describes an abundance of hills, depressions, stony outcrops, naked rock mountain ranges, and the most prominent feature: the enormous alkaline salt pan, a vast flat area of crusted ground commonly referred to as the playa.

The Black Rock playa is high desert, 3,848 feet or so above sea level, and the locals say it is the largest expanse of uninterrupted flatness in North America, save possibly the Bonneville Salt Flats in Utah. It is shaped like a distorted wishbone and is divided into three parts: the stem, the west arm, and the east arm.

Black Rock also has a large concentration of hot springs, mostly in the border zones where the mountain ranges meet the playa. The source of the springs, which have names such as Trego, Soldiers, Black Rock, and Double, is still unknown, although it is theorized that they are the result of the active volcanism and latent heat from the Cascades in northern California and Oregon.

The route takes you east from Reno on Interstate 80 to the town of Wadsworth, where you pick up State Highway 447 northbound. After a few miles, you encounter Nixon, Nevada, and then it's a brain-on-autopilot streak through the desert 120 miles due north to Gerlach and, a little beyond, your final destination: Black Rock City.

You and your co-pilot drive on, headlights on now in the gathering darkness, and after Gerlach you make your way the remaining 10 miles onto the billiard-table-like flatness of the playa. After making a turn, there, in the distance, you get your first glance at Black Rock City. It is amazingly bright out here, in the middle of nowhere. You can see people riding about in cars and buses that look like pirate ships, giant ducks, mechanical birds, and Pakistani jitneys. It's a bit dark, but it appears to you that many of the women riding past you on bicycles are pedaling through the desert evening without tops, and in some cases without bottoms either. There's techno music coming from one side and shouts coming from another. Some people are dressed in outlandish costumes and some are barely dressed at all.

"Welcome home," says the greeter with enthusiasm at the gate. "Have you been here before? Yes? No? Well, then, glad to see you." The greeter gives you the rundown on what behavior is allowed and what's not allowed.

"Black Rock City is your community," she says. "That means that while you're here, this is your place and it's your event. You are a citizen and are responsible for making it a good place to be! Participate—this place has no spectators; you are the event. Most things are allowed except for firearms, fireworks, dogs, bad attitudes, overly inhibited behavior, and doing things that harm the playa or your neighbors. See ya. Peace out."

You are now in Black Rock City, or BRC as it's normally shortened to, and it has literally sprung up in the last week from a population of none to this horde of humanity. Almost overnight this has turned into Nevada's fourth largest metropolis.

> > >

Welcome to Burning Man, the extravagant, hedonistic, riotous countercultural festival that takes place deep in the Nevada desert every year in the week preceding Labor Day. Thousands of people make this annual pilgrimage to the hot, dry Black Rock Wilderness to mix with like-minded natives in the temporary society called Black Rock City. The city is built on the Black Rock playa, one of the flattest and most barren parcels of land in the Western Hemisphere. There are no shrubs, trees, rocks, animals, or topographical elevations of any kind. It is simply a vast expanse of sun-baked flatness, normally totally devoid of life . . . but for a couple of weeks at the end of each summer.

A common public misconception is that Burning Man is a week-long rave, nothing but drugs, sex, and rock and roll—something like an ultra-marathon Grateful Dead concert. In truth, it is a celebration of temporary community building and self-expression. The purpose of attending the festival varies between individuals. Some seek meaning in a New Age sort of way, some come to be part of large gathering of like-minded people, and some come to party to excess. The festival culminates in the ritualized burning of a large wooden sculpture of a man. The meaning and symbolism in the act of burning are left to the beholder to interpret.

Like San Francisco and its Silicon Valley, where, believe it or not, this concept got its start, Burning Man is a technology incubator of outlandish proportions. Silicon Valley technology workers become creatively inspired by the 180-degree opposite that the desert provides from their normal work environments. It seems that Black Rock City provides a near-perfect outlet for the creative urges normally kept under wraps at home and at work. This is a place for sharing ideas and concepts about technology that entertains. All sorts of technical types come here—robot builders, pulsejetters, Tesla coilers, laser enthusiasts, and machine artists—to commingle, work together, exchange ideas, talk shop, and show off their best stuff.

There has been a fair amount of thought invested in describing and understanding gatherings such as Burning Man. In Australia, there are similar gatherings held in the outback. There, such encampments are called "doofs." One Australian social-sciences academic who has the amusing job of going from doof to doof and studying them describes these events as "both a selfless and . . . a self-indulgent counter culture, fusing social critique with abandonment and escape to the dance beat, to pleasure."[1] Burners, likewise, are simply the biggest subset of a culture of exotic, exhibitionist thinkers, technicians, and artists who wish to share their ideas with whoever comes by, without censorship.

INSIDE BLACK ROCK CITY

Burning Man's Black Rock City covers well over 3,000 acres. The event's planners have arranged the 5 square miles of BRC on a circular plat, with the streets laid out in concentric rings radiating outward from the iconic wooden statue of a man, the Burning Man effigy, in the center. Each street is lined with tents, converted car campers, recreational vehicles, and a wide variety of ad hoc structures, some large enough to hold scores of people. The shelters vary widely in construction materials and methods. Some are big tents. Many are geodesic domes. And others are just frameworks made from plastic or iron pipe, frequently covered with camouflage netting or surplus parachutes so as to provide large areas of shade.

The innermost street, the one closest to the Burning Man effigy, is the widest pathway and is called the Esplanade. This street is fronted with the largest and most elaborate structures, and is filled with an around-the-clock assemblage of walkers and bikers. The structures on the Esplanade typically encompass multiple tents or recreational vehicles with common areas constructed by large groups with common interests; these compounds usually share a sort of loose architectural theme, and so they're called "theme villages."

Some of the theme villages are simple in both design and execution, but many appear to be the product of rather prodigious amounts of labor and materials, considering the mayfly-like lifespan of the city. It is not unusual to come across encampments with 26-to-30-foot-high wooden towers, including elaborate parapets and crenellations, yet still wrapped in the ubiquitous camo-colored cloth covering, making the area look medieval and post-apocalyptic, part King Arthur and part Mad Max.

THE VIEW FROM ABOVE: KITE AERIAL PHOTOGRAPHY

During the hot afternoon, a man wearing a too-big fedora and sandals walks down the Esplanade, flying a kite over the playa. He is simultaneously reeling out his kite and manipulating some radio controls similar to a transmitter used to control the servo motors on a robot.

A group of women walks up to see what's going on. "So," says a short, stocky young woman wearing an outfit of her own design that exposes most of one breast, "what's up with that thing?"

"It's a kite I made to take aerial photographs," he says. The group around him seems interested, so he continues. "I've gone all over the world taking pictures from the sky. Here, take a look." He reels it in to give the group a closer look at the kite, a two-planed double-decker affair.

"Did you invent this?" asks one of the group.

"Oh, no. It may have been used in Europe as far back as the 1880s," says the kite man. "I know for certain that it was used at the time of the San Francisco earthquake, because I've seen the pictures it took." The kite man lets out some more line, and the kite ascends in the strong, dust-laden breeze.

When airborne, his kite typically flies about 75 feet above the playa. It is designed such that a digital camera hangs about 20 feet below the kite tail, suspended in a special harness he designed.

The harness contains two small radio-frequency position controllers called servomotors. By radio control he can aim and position the direction and focus of the camera lens from the ground. When he carefully manipulates the joysticks on his radio controller, the kite cameraman can capture an aerial snapshot. Kite Man can get a 360-degree view from high above without leaving the ground. He displays a few samples of his pictures for the assemblage.

"The best part of this thing is the incredible hawk's-eye perspective. I've been all over the world with my kite setup, and I've got pictures and viewpoints that no human has ever seen before. I've got treetop views of the Great Pyramid, and there are no trees there, of course; I have taken pictures from angles that no person has ever been able to look from. I've got pictures of the Grand Canyon that show details and panoramas that no climber, no matter how talented, can ever match. Just me and the birds have seen some of this."

The kite cam man continues on down the Esplanade taking pictures. He is frequently stopped and questioned by the curious. People virtually line up to talk with him, and he never tires of describing what he is doing and showing off his photos.

Much later that night, Kite Man is still walking the playa, still flying, still meeting people. "My kite cam," he says with a big grin, "is the best thing inside Black Rock City for meeting women."

THE BLACK ROCK RANGERS

Everybody who comes to Black Rock City must have a ticket to enter, and tickets are not cheap: from about $150 to $300, depending on when they're purchased. It takes a pretty expensive ticket to ride this party train. There are people—many people—who would prefer to make a cross-desert run for the perimeter fence in the dark of night than part with their money. So a group of volunteers called the Black Rock Rangers watches the edges of the city all day

THE TECHNOLOGY OF
KITE AERIAL PHOTOGRAPHY

The French photographer Arthur Batut took the first kite aerial photographs over Labruguière, France, in the late 1880s. He developed a primitive camera cradle and shutter activation system. The camera was supported from the kite itself, unlike modern kite camera rigs, where the camera is supported from the kite line. Batut devised a slow-burning fuse that activated the shutter a few minutes after the kite was launched. When the shutter released, a small cloth fell to earth, signaling Batut that it was time to haul the kite down. Kite aerial photography has evolved considerably since then.

In the early 1900s, a midwestern American photographer named George Lawrence suspended the camera from his kite line. Separating the camera from the kite reduced camera motion and resulted in sharper pictures. In those days, cameras were big, and therefore so were the kite rigs that floated them. Lawrence mounted a 46-pound camera to an array of sixteen box kites and took aerial photographs of post-earthquake San Francisco. His spectacular photos are the only ones that are able to show the true scale of destruction.

In 1912, another Frenchman, Pierre Picavet, invented the now almost standard cross-shaped kite camera suspension system named after him. The Picavet mount makes use of a rigid metal cross-shaped spacer suspended from the kite line. Each of the four ends of the cross is connected to two attachment points looped into the kite line.

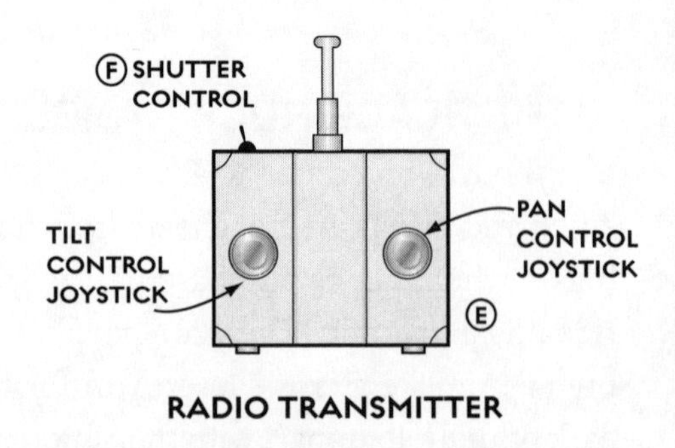

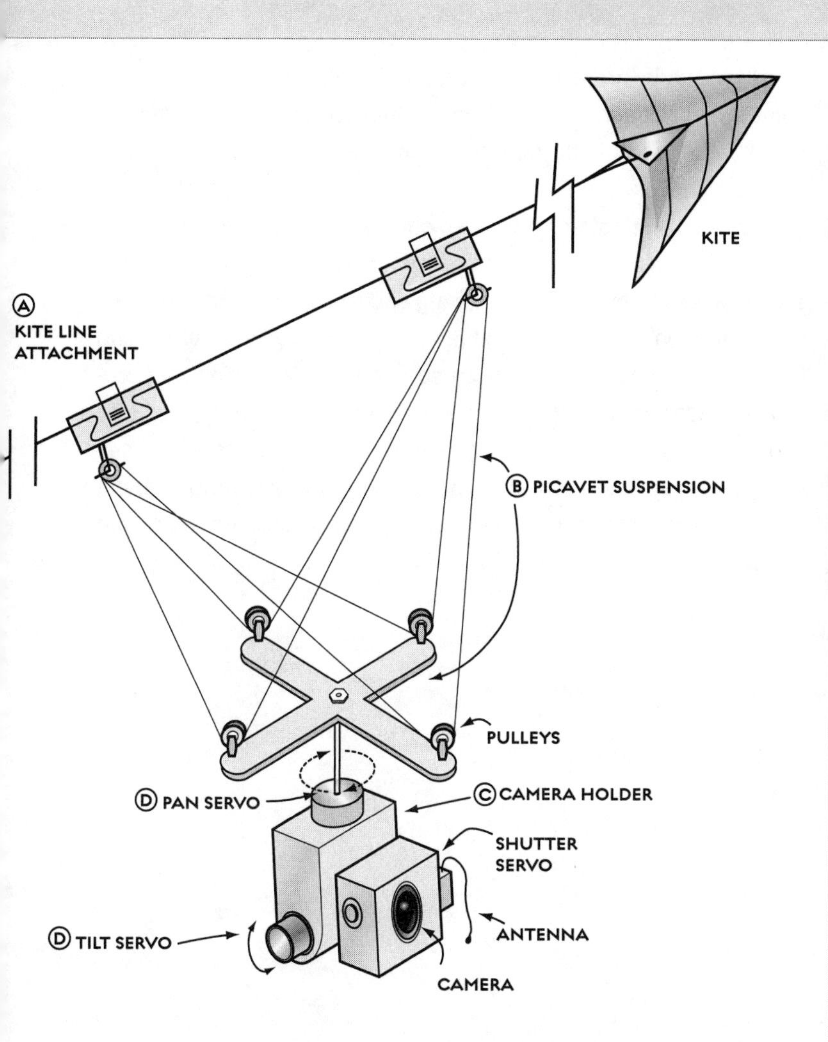

1. The kite aerial photography (KAP) rig is attached to the kite by two attachments **(A).**

2. The camera is kept level, irrespective of the kite line's angle to the ground, via the Picavet suspension **(B).**

3. The camera is held in a holder **(C)** outfitted with two remote-controlled servomotors **(D).**

4. One servo controls the camera's tilt, the other controls the pan.

5. The photographer controls the servos through a handheld radio transmitter **(E).**

6. The transmitter has a shutter release control **(F).**

Often, pulleys are used to attach the Picavet to the line. By using a long continuous line with multiple attachments, Picavet's system damps out the buffeting and twisting of the kite. Pierre Picavet's system provides two advantages: First, it keeps the camera's spatial orientation somewhat constant. The camera holder, or cradle, will maintain its preset horizontal attitude regardless of the angle formed between the kite string and the ground. Also, the two cable attachments spaced several feet apart on the kite string keep the camera fairly well oriented in a single direction and prevent the camera rig from rotating wildly in the breeze.

Second, the cat's cradle of camera-holding suspension lines damps the up-and-down motion of the camera. Without them, there is a lot of random and sometimes violent pitching and jerking. <

and night with the best homemade and underground-style security technology that money can't buy, and they try hard to maintain BRC's perimeter integrity.

The Black Rock Rangers have developed some pretty interesting guerrilla technology to keep the freeloaders out. For one, there is a large radar tower in the middle of the main festival area, which is called Center Camp. From the top of a tall, spindly tower, an old Furuno radio constantly scans the whole area with its all-seeing, rapidly rotating, radar-frequency eye. The radar was scrounged from an old marine radar setup, but has been modified by the Rangers so it can spot any vehicle from a bicycle to a semi that is out on the playa in places where it shouldn't be. Coupled with a patched-together multichannel radio system, night-vision tools, and desert-worthy 4 × 4's, the Rangers figure they have the technology to handle whoever and whatever tries to sneak past them.

THE BURNING MAN

The centerpiece of the festival is, as it always has been, the wooden effigy of a man, arms outstretched, lozenge-shaped head looking out over the desert in the center of the camp. This is the Man, the symbol of everything that Burning Man is about. The festival's highlight comes on Saturday night, when the 40-foot-tall man is set alight, creating a magical atmosphere in the dark desert sky.

The Burning Man effigy has a peculiar sense of proportion that is naturally appealing. He looks much like Leonardo da Vinci's Vitruvian Man, the subject of the famous pencil drawing of a naked man with leonine hair and outstretched arms. The drawing of the Vitruvian Man shows a square inscribed inside a circle. In it is a man with outstretched arms and legs. More correctly, there are two pairs of each, and the extremities touch both the circumference of the circle and the vertices of the square.

It is not just that the Burning Man effigy mimics the Renaissance-era Vitruvian Man. Many similar combinations of old and new abound. At the burn ceremony, there is an appealing and yet unsettling juxtaposition of the primal with the high-tech. When the Man burns, he does so with exceptionally great vigor, partially due to the natural and ancient tendency of desert-air-desiccated wooden structures to burn with a lightning-quick release of energy, but also partially resulting from a ton of paraffin oil accelerant soaking the wood, electronically controlled fireworks, and a battery of liquefied natural gas flame cannons.

When the Man burns, the highly choreographed show begins with fire dancers and moves on to fireworks. The Man is outlined with neon-like lights, and next to him, great rising circles of fire roar from propane cannons placed near the base. Red, green, and purple lasers of enormous size light the sky. Then the Man burns, and he does so spectacularly, in a great roiling ocean of chemically accelerated flame. Many people, even those who have come for years, aren't sure what it all is supposed to mean.

But the answer is simple: It means whatever the observer wants it to mean.

Freedom, anarchy, fire, and dangerous machinery have been part of the Burning Man experience since it began on a beach near San Francisco Bay in the early 1990s. As the festival grew, the location moved to the Black Rock Desert in order to accommodate the increasing number of people. Now it has grown (some say metastasized) into the largest experiment in temporary community building in the world.

Veteran Burners often speak longingly of the "old days," which, if popular accounts are to be believed, were far more anarchic than the present events. For in the mid-1990s heyday of Burning Man anarchy, the absence of rules and restrictions of almost any sort gave the event not just an aura but an actual and tangible danger, as well as a sense of true freedom. Burners could do pretty much what they felt like—there was no one laying down rules regarding the use of fire, firecrackers, firearms, machines, pyrotechnics, flamethrowers, or anything else.

Back in the early days, it was not unheard of for metal-winged art cars to rev up to 75 miles an hour and zoom down the playa without headlights at midnight. There was no authority setting those types of limits. Such freedom and its attendant and evil step-sister, irresponsibility, took their toll.

In 1996 a couple of people were run over in their tents, and another person was killed in a collision while riding his motorcycle out on the playa. In a chaotic frenzy that occurred after the Man burned, too many people lost themselves in a self-induced state of fire lust and torched everything that would burn—including other people's property. The event had to change or the whole thing would die.

The next year, eventgoers could no longer drive on the playa, wildly or not. The places where fire art could be displayed were regulated, and fireworks were banned entirely. The admission ticket, which used to read "Please keep weapons unloaded in camp," now said flatly "No weapons, period." And, as the Burning Man event staff took necessary steps to deal with the growth and enlarged scope of the festival, the whole thing changed: With the anarchy and the anything-allowed attitude went the complete freedom of self-expression. Some still mourn its passing.

SATAN'S **CALLIOPE**

On a wide-open section of alkaline salt pan, a few hundred yards away from the Man, stands what on a per-cubic-foot basis is probably the world's loudest thing. Cordoned off behind two fire-breathing dragonheads works the "Engineer from Hell," Lucy Hosking. Lucy tweaks and tunes her invention, the painfully loud pipe organ made from jet engines that she calls Satan's Calliope.

This is a fire-breathing pipe organ mounted on an electric car. The base vehicle is a barely recognizable 1989 Harley-Davidson electric golf cart outfitted with a body made of aluminum fish scales. The fire-breathing organ is mounted to the rear of the car and to hydraulic cylinders so that it can be made to lie down for travel and then rise to a commanding vertical orientation for performance. It is a heavy beast, approaching a ton with a driver inside.

Satan's Calliope is a jet-assisted organ with several different voices, or organ sounds, all of which are driven by metal tubes made to vibrate by moving gas, just as in true pipe organ fashion. However, this organ is sounded by a mixture of different types of pipes, including a bank of seventeen pulsejets, fourteen truck horns, fourteen air-driven pipes, and, just for the heck of it, two whooshing gas flares. Turning on and off the pulsejets is like starting and turning off a car engine. When Lucy presses the organ key the sound may start instantly or it may take five seconds. So, while it is possible to play chords, anything with a rhythm or a beat is nearly impossible.

The pulsejets form the organ's bass voice, and the regular air pipes are the treble. The truck horns double up in the range of the pulsejets' bass, but they screech often enough and are capable of making some really high notes. Topping it all off are the two big accumulator flares that make a *foooom* sound.

The loudest noises made are single pulses from the pulsejets; this is much like an automobile's backfire or a truck using an illegal Jake Brake.*

---

*The term "Jake Brake" refers to the engine brake manufactured by Jacobs Vehicle Systems of Bloomfield, Connecticut, for large over-the-road trucks. The Jake Brake works by altering the action of the exhaust valves, thereby turning the engine into a giant air compressor. Installing a Jake Brake has the effect of converting a power-producing diesel engine into a power-absorbing energy sink. This adds considerable braking power while reducing wear and tear on the truck's normal hydraulic brakes.

The Calliope's fuel is carried in two 15-gallon horizontal tanks located under the vehicle's side panels. An array of elaborate pressure systems maintains safety during organ performances.

The pulsejets use compressed air in addition to propane. The air is primarily used to start the engine, and it is then shut off by an onboard computer after flame ignition. Lucy can create different special effects by changing the timing of the fuel, air, and ignition to the jets. Buzzes, rumbles, and explosions are all possible.

Besides possibly being the world's loudest thing, Satan's Calliope probably lays a more specific claim: the world's most elaborate pulsejet. This is a nontrivial accolade because pulsejets occupy a very special place in the panoply of Underground technology.

At nightfall, Lucy Hosking drove Satan's Calliope out onto the playa, almost underneath the Burning Man effigy itself, for a concert. The pulsejets operate at very, very high temperatures, and there are concomitant dangers involved when using compressed propane. So no one can get too close to it when it is operating, and therefore Hosking plays her machine from a distance, using a detached guitar-style keyboard with a shoulder strap.

Even through hands cupped tightly around ears, the sound was deafening. Says Lucy, "Any sounds coming out of Satan's Calliope that resemble Western music are an unfortunate accident."

CAMP **PUMP**

Camp Pump, a large, white, round tent with a baffled tent flap opening, was not a bit unusual from the outside, but inside things

---

It has one big drawback: It is very noisy. You likely have heard a semi use its Jake Brake without realizing what it was. Sometimes, when a truck is approaching a stop sign, you might suddenly hear a loud roar, like a very large lawn mower revving, for five or ten seconds. That roar is the Jake Brake slowing the truck. The compression noise associated with an active Jake Brake has caused many U.S. cities to pass ordinances against their use.

THE TECHNOLOGY OF PULSEJETS

Pulsejets are very simple, resonant jet engines. If built well, they are dependable and provide considerable amounts of thrust. Unlike most jet engines, which require highly specialized alloys and expensively machined rotating parts—including high-speed rotating turbines, fans, nozzles, and fuel injectors—a pulsejet is easily made in the tinkerer's garage by anyone with a measure of fabrication skills. Therefore, pulsejets are coveted in the Underground. Pulsejets, say their admirers, are terrific little items, almost magical—basically just a stovepipe with a few carefully placed holes. But what a powerful and loud stovepipe!

Pulsejets, like other engine types, operate on what engineers call a "combustion cycle." In the first part of the cycle, air is sucked into the combustion chamber. At the same time, fuel, in the form of a hydrocarbon gas (for example, propane), is also introduced into the combustion chamber and mixed with the air. In the next part of the cycle, a spark or other ignition source

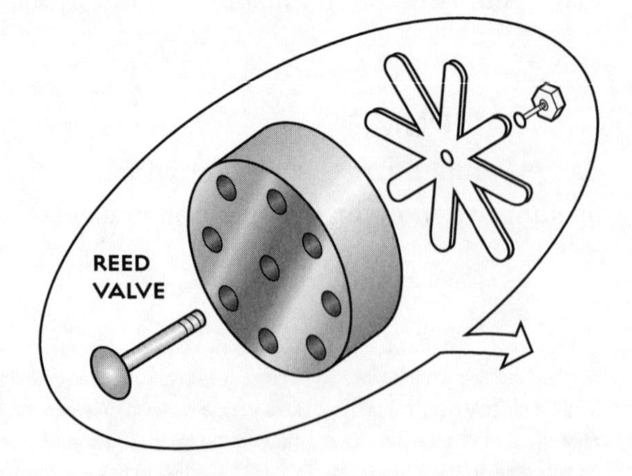

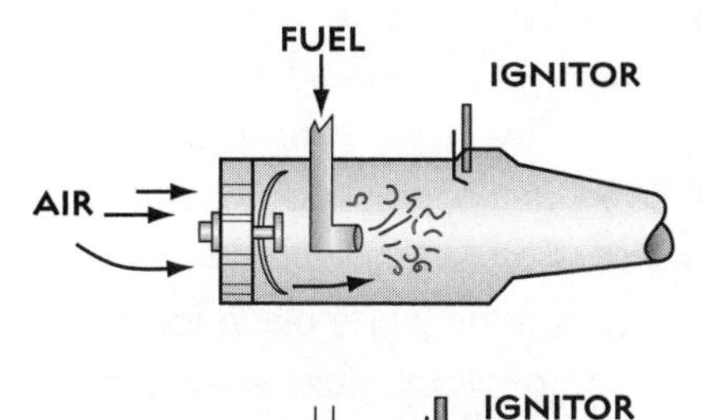

1. Reed valve open, air is taken in, fuel is taken in.

2. Fuel/air mix ignites, pushes reed valve closed. Thrust results. Exhaust leaves, creating vacuum. Reed valve opens, cycle repeats.

3. Pulsejets have one moving part: the reed valve.

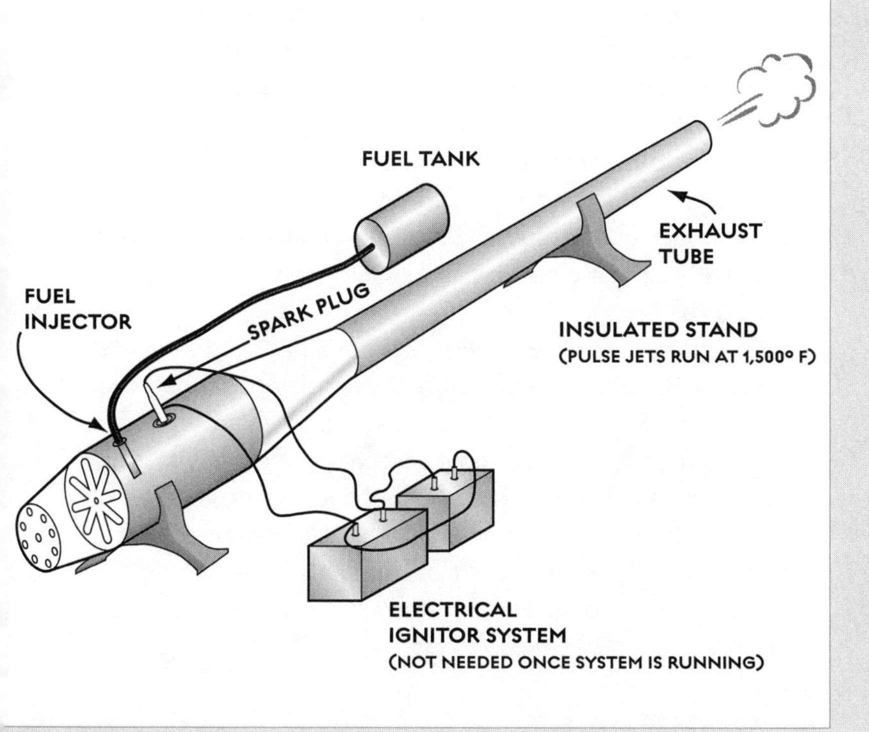

ignites the fuel-air mixture, and *kaboom*—the expanded gas rushes out of the nozzle, resulting in the same forward momentum and pressure thrusts described earlier in the section on rocket engines.

A pulsejet contains a spring-loaded shutter-type valve placed ahead of the combustion section. Air is admitted through the valves, and when combustion is initiated, the pressure increases. The pressure increase in the combustion chamber presses the one-way intake valves closed, thereby preventing backflow through the inlet. Now there is only one place for the hot gases to go: They are expelled through the nozzle, producing forward thrust.

After the fuel and air rush out the nozzle, the combustion chamber is suddenly evacuated of fuel, air, everything—it all went whooshing out the nozzle. So now, momentarily, the air pressure within the combustion chamber is lower than the surrounding atmosphere. Since nature abhors a vacuum, the third part of the cycle ensues—air rushes back into the combustion chamber through the spring-loaded one-way valves placed ahead of the combustion chamber (that is, in the end opposite the nozzle), and the process is ready to start over. In a pulsejet, a self-sustaining suck-boom-push-suck cycle occurs within the combustion chamber thousands of times a minute, resulting in the oh-so-loud and annoying buzz that qualifies the pulsejet as one of the loudest things on earth for its size.

One great thing about a pulsejet, from a backyard tinkerer's viewpoint, is that the pulsejet engine's basic design has only one moving part, the one-way check valve on the air intake. Thus it can be constructed by just about anyone with the ability to weld ferrous pipe and mill slightly complex shapes out of sheet steel.

Adding to its backyard enthusiast appeal, the true mechanical simplicity of the engine means that the pulsejet does not require lubrication or a cooling system, although it must be made from materials that can withstand very high temperatures. Pulsejets run hot, the casing heating up to cherry red, signifying an operating temperature around 1,500°F.

This may be oversimplifying matters a bit, because making a pulsejet requires some careful machining, drilling, and a measure of welding. The hardest part is making the pulsejet's valve, which is usually fashioned from machined steel plate cut in the shape of a flap or daisy and fixed so that the petals fit over the intake holes. Most engineers would term this a "reed valve."

Given that pulsejets are so easy and cheap to build, why is it that they have almost no current commercial use? Well, there are three drawbacks. They are fuel-inefficient—that is, they consume an awful lot of gas for the thrust provided. Second, the vibrations produced from the pulsing action are so great that no one could ride a vehicle powered by one. Finally, they are too noisy to use commercially anyplace north of Antarctica.*

*Pulsejets are simple, they are powerful, they glow red hot when operating, and, most of all, they are loud. The loudness of a pulsejet is its most distinguishing characteristic, that and its edgy reputation.

Traditionally, pulsejets carry with them a vague evilness, a sort of Third Reich type of vibe. This is due to the device's military and German heritage. It was during the early 1930s that a German engineer named Paul Schmidt first developed the idea for the pulsejet engine, but it didn't immediately catch on. During World War II, Germany's military planners dusted off the idea and started production of the V-1 rocket (the *V* was short for *Vergeltungswaffen,* which roughly translates as "weapon of retaliation"). The V-1 was simply a winged pulsejet outfitted with nearly a ton of high explosive as payload. The Germans made and fired thousands of these terror weapons—unmanned jets, really—and they were extremely destructive.

The distinctive, resonating, and incredibly loud sound of the V-1 pulsejet resulted in its nickname, the "buzz bomb," coined by the English. People on the ground knew they were probably not in imminent danger if the buzzing sound came and then slowly faded as the buzz bomb passed over. But if the buzzing suddenly stopped, they'd be in big trouble, as that meant the V-1 had run out of fuel and was on its way down with 2,000 pounds of explosives fitted to an impact trigger.

Every single V-1 ever launched was nominally aimed at Tower Bridge in London, but because the aiming controls were so primitive, they basically fell all over southeastern England. There were close to seven thousand English buzz bomb casualties before the Allies overran the launch sites in France and Flanders to stop the assault. <

THE TECHNOLOGY OF VACUUM ENHANCEMENT

When Nature Adores A Vacuum

Vacuum devices have been prescribed by urologists since the late 1970s when they were first recognized as a valid alternative for men with erectile dysfunction. Now there are many vacuum products out on the market, not just oddball sexual devices, but products that are actually FDA approved for just this application. There is a compelling economic proposition to vacuum devices as well—the cost of a vacuum erection device is about $150 and the cost of a prosthesis with surgery is $8,000 or more.

The vacuum device works by reducing ambient pressure around the penis. The user's member is coated with a generous amount of lubricant and then inserted into an air evacuation cylinder. The flanged base of the cylinder is placed over the pubis with another generous dollop of lube to act as a seal. Once the desired erection is produced, a medical-grade rubber band is transferred from the base of the cylinder to the base of the penis. The result is a water bag phenomenon—like taking a latex glove, filling it up with water, and twisting closed the wrist portion of the glove.

The vacuum-initiated erection differs slightly from a normal erection in that the root of the penis, that is, the part below the rubber band, is somewhat softer than the portion beyond the band. One urologist tells his patients that what they'll have is a loveable, stuffable penis.

More recently, vacuum devices have been sold with the promise of increasing length and girth. Unfortunately for those who daydream about such things, the mainstream medical literature does not have much positive evidence regarding the efficacy of this application.

The Camp Pump patrons and others who chronically use a vacuum device (a practice that, for the record, is often not recommended by medical doctors who study such things) frequently experience what urologists colloquially term "bronzing of the skin." This is caused by the rupture of small capillaries that leaves a deposit of iron pigment (from the red blood cells) under the skin. The penis will become congested and appear larger (to some, the word *swollen* may be more appropriate). However, when users take a holiday from using the vacuum, their penises revert back to normal. <

were pretty weird. Camp Pump's entry into the radical technology sweepstakes consisted mainly of a collection of large Lucite bell jars and vacuum pumps. Inside the confines of the tent there were often as many as half a dozen naked men sitting on lawn chairs, each with a clear plastic jar firmly clamped over his groin. Each was busily operating the lever end of an attached vacuum pump, which was slowly but surely inflating their testicles to the size of oranges.

The man in charge of the operation sported a bizarrely enlarged scrotum. He appeared very proud of his package and was not at all self-conscious, although he had a sack big enough to double as a saddlebag. Upon entering the area, a straight-talking newcomer sized up the host with wide eyes and exclaimed, "Dude, you've got some wicked big nuts there. What's it feel like?"

The man with the large testicles was quite willing to explain his hobby, for he knew much about vacuum technology and the biology of body-part pumping, and he could discuss his passion in very matter-of-fact terms, much like, say, the barista at Starbucks explaining how her espresso maker works.

"I'm not sure about what level of vacuum is drawn within the bell jar, but if you pump long enough I think it gets pretty low. The vacuum level is significant."

While he was explaining the physics and biology behind the activity, a young skinny guy with a large floppy hat and plenty of body art waved at the owner. "I'm having trouble keeping the suction going," he said. "The air leaks in too fast."

The proprietor nodded. "Yeah, it's sometimes real difficult to maintain a good seal, especially as hairy as you are. Put on some more lube."

He continued, "Here, I'll demonstrate how it works. To use it, you just insert your member in here, and then draw a tight seal around the scrotum." He demonstrated on himself. "Use lots of lube. Believe me, lots of lube is the key."

He made a squishing sound as he adjusted the jar for a tight seal.

"And be sure to take frequent breaks during the pump up. What you're doing is moving blood into the fleshy, air-filled areas of your body, the area underneath the glass. It takes a while to expand those parts, which is why it takes an hour or two to get good results."

After enlarging for some time, the pumpers released the vacuum on themselves and bade Camp Pump farewell. They left through the tent flap and walked out into the desert, their newly inflated testicles exposed for public view. As they no doubt had hoped, they gathered stares quite frequently.

DR. MEGAVOLT

A man in an iron mask starts to dance on top of a recreational vehicle on the Esplanade. It's well after midnight. Up there with him, on the flat roof of the camper, is a large metal doughnut, the size of a semi's inner tube, perched on a wire pole with a sign reading MUYTATOR and DANGER! HIGH VOLTAGE. There are a couple of thick electrical cables hanging down the backside of the RV.

Suddenly, a display of intensely bright lightning bolts, 12 feet long and composed of hundreds of thin electrical fingers, flashes in the darkness. Electrically induced cracks and snapping soar out over the heavily amplified techno music booming across the festival city.

The Dr. MegaVolt experience is part science lab, part midnight rave: a giant Tesla coil that provides entertainment to large gatherings of Burners from atop his recreational vehicle.

A typical Dr. MegaVolt show begins sometime after midnight. First, Austin Richards (aka Dr. MegaVolt) and his helpers start the gasoline engines that power the oversized portable electrical power generators required to provide a sizeable quantity of power miles from the nearest power line. The hum of the generators is a signal that the show will soon begin. The techno beat of music surges, and large crowds gather around Richards's RV.

After a suitable interval to build tension, he dons a homemade chain-mail suit—electrical armor—and takes his position at the base of the large and powerful-looking coil. Richards's suit is made from specially engineered loose wire mesh designed to protect him from the six-digit voltages that would stop the heart of an unprotected human. The suit exploits an electrical phenomenon called the "skin effect" and the wires form a type of Faraday cage around his body. The suit protects him by maintaining a safe level of electrical potential from his head to his feet. The electricity doesn't penetrate the suit; it merely sloshes around it like water against a neoprene wet suit. The electricity travels around the periphery, leaving the wearer untouched, electrically high and dry. Without the suit he'd be a smoking pile of cooked meat, charbroiled to well done, without a warm pink center.

"Working with high-voltage is not for the untrained," he says. "I've gotten shocked a few times and it's no fun. I tried to light a cigar on the arc of a coil and it didn't work out—I shocked the inside of my mouth. That's a bad place to get a shock. A really bad place.

"It's taken a while to build up the knowledge I have in order to do this without killing myself. I built the coil we use now back in 1991 at the University of California at Berkeley."

Over time, Dr. MegaVolt has aimed higher and higher. "You know, by themselves, the coils are pretty interesting. But when you get down to it, it is just technology—beautiful, but lifeless. There was something missing: a human element. So in 1996 I started to collaborate with members of San Francisco's machine artist community. We built a cage to protect a person from the high Tesla coil currents.

"As a stunt, I got inside the cage at a party. The results were exactly as I expected: I felt no sensation of electricity while in the cage. All the current flows on the outside of a conductor, isolating the interior from the dangerous electric fields. I thought about this and determined that I could shrink down that protective cage into a metal suit that would allow me to walk around."

As he walks the top of the stage built on the roof of an RV, the ultra-high-voltage electricity zips and swirls around him, and great crackling arcs leap to his hands from the toroid-shaped electrode at the top of the secondary inductor that towers high over the playa. He looks like Zeus, throwing and catching lightning bolts from his fingertips and bending the arcs of ionized plasma into shapes to suit his mood.

The spectacle produces a storm of excitement. Generally, by this late in the week it's hard to faze the Burners, most of whom are by now inured to giant art, noise, and flashing lights—they see it as a natural part of their environment. But Dr. M. still captivates them. The primeval lightning, enabled by early-twentieth-century electrical constructions and presented in a postmodern artistic setting, makes it impossible not to enjoy.

Richards has been doing this stuff for years. He saw his first coil at age eight, back in his hometown of Newton, Massachusetts, and he and his friends were soon building spark-gap devices for stage props. His stuff was the hit of the Halloween shows staged for the neighborhood kids.

The whole mad-scientist idea resonated with him as a young man. After completing high school and undergraduate work, and getting a doctorate in physics at UC Berkeley, he still thought about coils. He worked at the Lawrence Berkeley National Laboratory near San Francisco and spent several summers on subatomic particle experiments at the Amundsen-Scott South Pole station in Antarctica. Then he came back to coil building, the bigger the better.

On the desert, the midnight crowd is mesmerized by the show. The lighting bolts continue to dance, and Dr. MegaVolt controls their movement. He shakes a metal trident and the electricity seems to obey.

To the surprise and terror of all present, one young woman, perhaps influenced by drugs, attempts to climb the RV and dance with the electricity herself. She takes a couple of steps toward the RV and then makes an effort to climb up. But she is quickly pulled down by others and carried away. Richards is angry, amazed that someone could be so stupid. But he calms himself and then the crowd. The show goes on.

3. TESLA coils

Some people spend their summers traveling between major-league ballparks. Some tour Civil War battlefields, and some visit Frank Lloyd Wright–designed buildings. But imagine that you've signed on for a rather specialized and esoteric cross-country tour, one that allows you to spend the next couple of weeks traveling parts of the USA, all of it in the southwest quadrant of the country. Your grand excursion has been dubbed "The Monsters of Voltage Tour" and you're a member of that group of scientific hobbyists interested in working with high-voltage electricity. On this tour, you plan to visit the best and most entertaining high-voltage machines and see them work. Your route takes you to visit the most energetic, historically significant, or largest arc-making, ozone-producing, ear-popping, high-voltage machines in the world. The first stop is Los Angeles' Griffith Park Observatory.

Most people huff and puff a bit when they decide to forgo their cars and make this trip on foot. It is a steep walk up the southern slope of

Mount Hollywood to the observatory building in Griffith Park, the largest city park in Los Angeles. The observatory has a commanding location on a high hill overlooking the LA basin. There are three large astronomical domes on top, one each rising from the middle, right, and left wings of the observatory. You've not been here before, but still the building seems a little familiar. Perhaps it is because it was the setting for many popular movies, including *Rebel Without a Cause, The Terminator, Dragnet,* and *Jurassic Park.**

Entering the main doors, you find yourself within the main rotunda. This is the reason you came up here—the world-famous Hall of Science. The hall is like a zoo for unusual scientific machine specimens. Its menagerie includes a triple-beam coelostat for observing the sun, a Foucault's pendulum ensconced in an altar of travertine, a rock from

---

*The Griffith Observatory, in which the world's most famous Tesla coil sits, is one of LA's most recognized buildings, a triple-domed building built in the Art Deco style. In 1912, Griffith J. Griffith offered almost a million dollars to the city fathers to build an astronomical observatory on land that he had previously donated and which had become Griffith Park.

The city refused the offer, principally because Colonel Griffith, as he was known, was a man of sullied reputation. (Griffith was always called "the Colonel," although there was scant proof he was actually in the military. One researcher suggests his military service was extremely limited. And evidence suggests that the only military title he ever held was major of riflery practice with the California National Guard.) In fact, the Colonel was a convicted felon. A few years earlier, Griffith had been released from San Quentin Prison after attempting in 1903 to murder his wife, Christina, popularly known as Tena, a wealthy socialite. Griffith, previously thought to be merely a wealthy, pompous blowhard, had gone off the deep end and shot the unfortunate Tena in the head. Disturbed, drunk, and delirious, he had got it into his head that Tena and Pope Leo XIII were attempting to conspire against him.

While vacationing at a Santa Monica hotel, Griffith had entered the room in which his wife was sitting. He had a religious book in one hand and a revolver in the other. He apparently handed Tena the prayer book, accused her of being in league with the Pope to conspire against him, and fired his revolver at her head. Tena jumped away at the last instant, sustaining a serious head wound but still managing to jump out of the nearby window. She somehow crawled to safety, and Colonel Griffith was arrested.

Griffith was convicted of attempted murder and spent a year or two in San Quentin. He returned from prison with his sanity restored and his pride humbled. He then made an offer to the city of Los Angeles to provide the money for the astronomical facility, but they declined. It wasn't until 1919, long after his death, that LA finally took his money and built the observatory. When it was built, it was built well, stylishly, and with sound scientific purpose.

Mars, meteorites of all kinds, telescopes, astronomical clocks, displays, and the Griffith Park Observatory Tesla coil. Above, the rotunda ceiling is painted with several large and intricate murals, symbolizing such sciences as geology and biology, mathematics and physics, astronomy, metallurgy, and electricity. The mural depicting electricity is especially relevant to this visit. It depicts Franklin's electrified kite, along with a man performing experiments with a spinning globe that's discharging high-voltage static electricity.

To the front is the large planetarium, with its ultra-precise Zeiss star projector. It provides a perfect reproduction of the night sky—any night sky, from any location, on any day of the year, and in any century past, present, or future—on the interior of the large domed theater. In the eastern section of the observatory's Hall of Science are displays of meteorites, seismographs, and astronomical clocks.

To the west is something else: something glowing and loud and demanding of attention. It is GPO-1, perhaps the most famous Tesla coil in the world. It is a noisy, bright, and beautiful manifestation of one of nature's most powerful natural phenomena—high-voltage electricity.

> > >

For a select few, the Griffith Park Observatory coil has proved life-changing. From its earliest days, this coil was said to be able to heal the sick, energize the weak, and enliven the brain.

In 1931, about a decade after the city fathers finally accepted Griffith's gift, the Tesla coil was presented to the city by Dr. Frederick Finch Strong, a licensed medical doctor with definitely non-traditional views. Dr. Strong was a confirmed believer in the purported restorative powers of high-frequency electrical fields. His treatments included placing his patients in close proximity to electrical fields of all kinds. Strong was a member of Los Angeles' Krotona Colony, a group founded by a number of East Coast occultists. The members of the Krotona Colony called themselves Theosophists and adhered to a philosophical pastiche of cabalistic spiritualism, occultism, Eastern religion, Masonic lore, and odd scientific viewpoints.

While ultimately electricity has become an important medical tool in items such as defibrillators and pacemakers, it was considered a compulsory tool by early Theosophist medical practitioners. Strong used electrical apparatus of various sorts to treat his southern California patients. After a period of experimentation, he came to the conclusion that what was necessary to improve the health of his patients was simply a more powerful and energetic electricity-producing machine. He commissioned an engineer to build a large, powerful induction coil, in the manner of inventor Nikola Tesla, who was a hero of a technological sort to the Theosophical movement. To fabricate the coil, Strong engaged the services of Earl Lewis Ovington, a former lab assistant to Thomas Edison.*

While there's no record as to the coil's pre–Griffith Park Observatory days, Dr. Strong likely used it in his Krotona Colony offices to treat patients for all sorts of maladies in the electro-medicinal manner during the 1920s. It was no longer used by 1931, perhaps due to changing perceptions regarding the safety and efficacy of such practices. Strong, however, understood and appreciated the general awe and amazement that his coil produced when switched on. So he gave it to the new Griffith Park Observatory's Hall of Science as a display. And there it stands even now, continuing to awe and amaze.

Griffith Park Observatory's Tesla coil has had an influence on science enthusiasts far out of proportion to its modest size. The GPO coil, although it commands attention, is not particularly extravagant in size or design. In fact, its compact overall design is the essence of Tesla coil simplicity. It consists of an elongated, truncated cone four feet high with turn after turn of tightly

---

*Ovington was also the very first airmail pilot in the United States. In 1911 he made a flight from Garden City, New York, to Mineola, New York, on official U.S. Post Office business, and delivered 640 letters and 1,280 postcards, including a letter to himself from the government designating him for all time as "Official Air Mail Pilot #1."

wrapped purple wire. The cone of wire rests on a squared-off wooden platform made from polished walnut. Rising from the flattened end of the cone is a 12-inch-diameter spherical electrode, spun from copper and sporting two pointed iron rods jutting out perpendicularly to the cone. A jumble of wires and boxes, with gauges and switches, connects these items.

The stylish coil's looks are the essence of yesterday's view of tomorrow: a sleek, modern, Art Deco style made from stacking the most basic shapes in solid geometry—a sphere on a cylinder on a pyramid—to create a sophisticated sculpture that happens to shoot lightning.

When power is applied to the GPO coil, it hums and crackles, sending out a brilliant shower of mini lightning bolts from the electrode toward the metal Faraday cage that surrounds it. Inside the workings of the coil, the output from a high-frequency, high-voltage transformer is amplified through the action of resonantly tuned electric and magnetic fields until the voltage exceeds a million volts at the electrode's pointed metal tip. The high voltage ionizes the air, breaking it down and forming a conductive path through the normally insulating atmosphere. The sparks create lucent, dendritic, purple fingers, forming and dissolving at such great speed that the human brain can only process and sense the composite afterimage that is retained by the retina. The bright filaments are made from nitrogen and oxygen atoms that have been converted into plasma. Plasma is a close cousin to fire (which a chemist would call a rapid oxidation process), but that's just an approximation of convenience.

When gaseous air, under the influence of high-voltage electricity, changes to plasma, the electrons of the gas atoms are knocked loose and the oxygen and nitrogen atoms exist temporarily as positively charged nuclei and negatively charged free electrons. These particles, or "ions," are the visible manifestations of the high-voltage electricity and are the electrical incandescence that is perceived as sparks and lightning.

The Griffith Park Observatory's Tesla coil is a device designed to produce very-high-voltage, very-high-frequency electric fields. When the coil is switched on, the field intensifies, and the air around the electrodes implanted in the copper sphere ionizes; the electrons are ripped free from their atomic centers and phase-change into conductive plasma. The first small streams of plasma at the electrodes form and then, as the electric field intensifies, more and bigger visible streams of light sprout, the plasma fingers lengthening into longer and longer streamers of bright, crackling, odoriferous light.

The plasma shooting from GPO-1 lengthens itself continuously as long as the distant end is still within the influence of the Tesla coil's electric field. The reason that the stream seems to continually lengthen is that the bright streamer has persistence. That is, it exists in the air long enough for the next dose of high-frequency voltage to add to it until it reaches its maximum length, which is determined by the coil's overall size. That's how the coil works—the bigger and more powerful the coil, the greater the coil's radius of effectiveness, and therefore the longer the streamers.

When the power to the coil is switched off, everything immediately stops, and the sphere on cylinder on pyramid on wooden platform sits quietly once more.

ELECTRIC CARPENTERS

Playing with high voltage is a surprisingly widespread pastime, given that it takes a special breed of individual to attempt million-volt electrical potentials in his workshop. There's nothing fragile or delicate about megavolt-sized experimentation. These people are the "rough carpenters" of the world of electricity.

There is a big difference between a cabinetmaker and a carpenter. A cabinetmaker works the wood, carving and drilling out bits in order to shape tight-fitting and secure joints. The cabinetmaker makes great use of a variety of files and precise, fine-toothed saw blades mounted on sturdy, rock-solid tables. His end product is fine furniture machined to tight tolerances.

The carpenter, on the other hand, uses heavy claw-handled hammers, steel crowbars, and long-bladed ripsaws to make tall-standing, sturdy, and strong structures. The carpenter makes garages, sheds, houses, and buildings—big and significant structures of all types.

There is an analogue to the carpenter/cabinetmaker comparison in the world of electronic hobbyists. There are people who love to tinker with highly precise, exacting, and sophisticated electric gear. These are people who can build their own oscilloscopes, for example. They are the electronic hobbyist's equivalent of a cabinetmaker. They know how to carefully test and troubleshoot solid-state electronic components and change the frequency and phase characteristics of audio recording equipment.

But then there are the electronic carpenters, those people who walk in the footsteps of pioneer scientist Nikola Tesla.* They work on big and powerful and dangerous electronic projects, mostly because they like big and powerful and dangerous stuff. Most of these people could, if asked, probably provide a reasonably good

---

*Few men have contributed so much to our world and impacted the rhythm and tenor of everyday life as inventor Nikola Tesla and received as little credit for their efforts (except from the high-voltage hobbyist community). He made enabling, fundamental contributions to so much of the technology that permeates modern life: radio communications, medical imaging, and most of all the AC electric motor, which opened the door to the widespread use of AC electrical power. Whenever a person plugs in a refrigerator, a sewing machine, or any electrical device with a motor, it works basically because Tesla figured out how to get power from the power plant to that person's home easily and inexpensively.

Nikola Tesla was born in 1856 in Smiljan, Austria-Hungary (present-day Croatia), where he grew up refined and intelligent, although prejudiced and obsessive-compulsive. It was apparent to all who knew him as a child that he was in possession of an incredible intellect and simultaneously plagued with myriad mental phobias and compulsions.

Tesla was often the subject of gossip due to a host of most unusual habits. He obsessively counted his steps when he walked, and he calculated the cubic volume of his food each time he ate. His fear of germs rivaled that of Howard Hughes. Human hair was especially repugnant to him.

He was very clean and neat, and for the record, tall and good-looking. A contemporary describes him graphically with this tongue-twistery flourish—he had "bushy black hair, brushed back briskly."

and intuitive explanation of how a capacitor works. More interestingly, though, they could explain how to make a big-time, lethally dangerous high-voltage capacitor in an unfinished basement, one able to store farads enough to stun a hippo; and they can make it from a twelve-pack of empty Rolling Rock beer bottles half wrapped in aluminum foil and immersed in brine.

One of the most intriguing characteristics of high-voltage tinkerers is how self-reliant they are. They make much of their equipment—capacitors, induction coils, switches, and so forth—from things they find in their garages and basements: coils of wire, copper tubes, PVC pipe. The high-voltage guys easily go from the tangible and practical to the electronic and invisible. Plus there's something exciting about setting up on a suburban driveway equipment that was built from scratch or found in junkyards and which sends up insanely intense electric fields.

There's a "big-science" connotation to high-voltage experiments, and that is something far removed from microcircuitry and doped silicon chips. High-voltage stuff is heavy, greasy, manly, powerful: iron and steel transformers, kilogram-sized ceramic

---

Tesla attended several schools and colleges, most notably the University of Prague, until he was banished for excessive gambling and "leading an irregular life." But while at Prague, Tesla started work on what would ultimately be his most important contribution: alternating current, or AC. Until Tesla, electrical power was distributed to homes and businesses in the form of direct current, or DC. Through the work of Thomas Edison and other electrical pioneers in the latter part of the nineteenth century, building and construction crews began to work on creating a web of copper lines carrying cheap, convenient electrical power in the form of DC to as many places as possible. For an electrical utility, though, DC has a lot of practical problems. On the plus side, it is easy and cheap to generate, because the machines—the dynamos and generators that produce it—are relatively cheap and simple to make. But getting that power from the generator to where it's used is another story. Power companies employing DC either had to use very thick (inches in diameter), heavy, and expensive copper transmission cables or build a DC generating station every mile or two. Neither alternative made electricity cheap or easy to get.

A combination of high voltage and low amperage, Tesla argued, was the key to cheap and ubiquitous distribution of power, and using alternating current was the only way to achieve it. But how to utilize high-voltage AC to do useful work was a difficult problem in

insulators, big four-gauge bare copper wire that's a full quarter inch in diameter—thick as a nightcrawler. Even the names of the things these electrical carpenters use have a macho quality to them: pole pigs, buss bars, tank circuits.

Best of all, high voltage is dangerous. Not just edgy, not just slightly hazardous to your health in the long term, but actually dangerous. Some of this technology is so risky to be around that the operating apparatus must be viewed from behind a polycarbonate or metal-shielded screen 15 feet away—and even so, people still risk some aspects of health and hearing. But the return is more than worth the investment: the ability to smell and hear molecules of air ionize into ozone and then see it explode into streams of multicolored plasma discharges, all produced by a homemade machine scrounged from stuff collected from eBay, basements, and junkyards.

John Dyer is a quintessential high-voltage hobbyist. His well-tuned and competently crafted coils work well and reliably. They are large enough to impress, yet small enough to become intimate with.

---

itself. Electric motors, by far the largest power-consuming component in industry and homes, operated only on direct current. AC power was considered unusable in motor applications.

It took Tesla more than four years to puzzle it out. The key to a workable AC power system, he finally discovered, involves the idea of rotation of conductors in and through magnetic fields. Multiple coils of conducting electrical wire, each rotating in a unique but repeating and unending pattern, would cause or "induce" alternating pulses of electrical power as they sliced through each other's magnetic fields at angles.

Electricity produced in rhythmic pulses, alternating between peaks of positive and negative electrical voltages, had the wonderful ability to be easily stepped up or down in voltage through simple electrical devices called transformers. Using stepped-up voltage, what formerly took tons of copper and steel to distribute could now take mere pounds. Due to Tesla's work, electricity became easier and less expensive to distribute and utilize. Tesla's concept of alternating current has proven to be one of the most important inventions in modern history.

Among the other great things that he accomplished, Nikola Tesla carved out a claim on the Extreme Tinkering Mount Rushmore for an invention technically called a high-frequency, high-voltage, air-core resonant transformer, but usually just called a Tesla coil. He developed this as a consequence of his experimentation with radio waves, and it is a favorite of radical technological self-expressives everywhere.

Dyer is a sometime radio engineer, electrical contractor, longtime Burning Man enthusiast, and Tesla coil devotee. John's neighborhood in Sacramento, California, is pretty tough-looking—kind of grimy and very industrial. His place of business, called Dyer Electric, is located in an unassuming part of town full of industrial and commercial space, empty lots, and high-barbed-wire-topped fences enclosing yards full of stuff that most people would have no interest in stealing, even without a fence.

His shop is a big warehouse-like affair with a group of voluminous work bays equipped with large garage doors and filled with recreational vehicles in various stages of repair. John has a large gray beard—a mountain-man-style beard—and thick glasses, often wears train engineer's overalls, and a friendly demeanor. Talkative and well-spoken, he greets all visitors warmly. His electrical hobbies have been important to him as long as he can remember. John's first word wasn't *mommy* or *daddy*. He said *plug*.

He repairs ancient recording-studio equipment, fronts a pirate radio station called K-RAP from an old trailer that he hauls to Burning Man to entertain the masses, and builds big Tesla coils.

"My religion," said John, "is the worship of the electron." That may be hyperbole or it may be true, but in any event John's house is certainly a shrine to the electronic arts.

In one of the many back rooms of his cluttered but expansive workshop, there are three or four separate compartments filled with all kinds of electronic stuff. The place is crammed to the gills with equipment, some of which is easily identified, and other items that are very mysterious. In the workshop, one hall leads to another, and in this labyrinth are dozens of parted-out reel-to-reel tape decks and audio mixing boards, vacuum tube power supplies, old condenser and electret microphones, a half-restored wooden boat, and a sizeable collection of old blowtorches. In the room farthest back stands a Tesla coil, one of several he's made.

The Tesla coil is usually considered to be the big cheese, the top gun, of amateur high-voltage electrical-amusement devices. It's the one that garners almost mystical respect from the people who build them. John's hobbyist-level coil is typical. The thing that most people first notice is typically the toroid, the topmost component, which looks like an aluminum Krispy Kreme mounted on a pedestal of coiled wire. This, in turn, is mounted on a steel box bulging with wires and electrical apparatus.

"Stand back," he warns during a demonstration, and dims the lights in the room. He plugs in his main transformer and then pushes the button that energizes the primary, or "tank," circuit. The coil energizes, and immediately sharp cracking and snapping fill the room. Thin blue lights, like the flame on a gas burner, start to appear. The noise grows louder and the arcs or streamers—"banjo strings," as John calls them—begin to lengthen, dancing between the upper doughnut coil and the surrounding air.

For illustrative purposes, John picks up an old, burned-out fluorescent bulb and holds it out from his body toward the coil. In his hand, the dead tube is transformed into a bright, glowing emitter of photons, resurrected by the coil's intense magnetic aura. In fact, all of the fluorescent lights in the room are glowing and blinking, even though all the power to them is off.

"Don't get too close!" he yells above the clatter. The streamers are growing longer and longer, feathering out from the toroid and dancing on air. They are beautiful, but a single touch could send your heart into arrhythmia or worse.

The arcs lengthen out, and the streamers start to double up and fan out, creating a dancing crown of visible electricity that encases the top of the coil. The popping and crackling continue to grow louder and louder until the sound becomes first overwhelming and then painful. The air smells bitter and metallic and the odor makes the nostrils burn and sting.

"The sting in your nose and throat is from the ozone. The coil is ionizing the air!" shouts John over the clatter. "It's probably not good for you to breathe." But there's no way to avoid it in the hypnotic presence of the live coil.

REALLY BIG COILS

Tesla coils are a viscerally interesting type of entertainment, and once the math and physics are understood, they are really not that hard to build.* Coilers live everywhere, and sometimes they meet in big groups to fire up their inventions and exchange tips and stories. But after a certain size is reached, the electrical demands on the components become very great, requiring special materials and building techniques, not to mention access points into the local power grid that won't cause blackouts when the thing is energized. There are only a handful of people in the world with the wherewithal and experience and space to build Tesla coils exceeding 25 kilowatts.

Jeff Parisse is a former rock and roll drummer who parlayed his experience staging musical events into the field of high-voltage special-effects presentations. Jeff started building coils in junior high and just never quit. Now he has a company called KVA Effects that builds high-voltage coils for all sorts of purposes, including permanent museum displays, movie special effects, and stage performances.

With his partners, he has designed and installed a host of coils both large and small. He has two favorite high-voltage machines. One is the coil he calls the SG-32, which is, depending on whom you ask, the third or fourth largest Tesla coil in the world. Under

---

*Tesla was also a spectacular showman, and his big electrical coils made for exciting entertainment. Although most modern experts now disagree, he claimed that the electric fields emanating from his Tesla coils were not dangerous because of their high frequency.

As the voltage at the top load or upper conductor of the Tesla coil rises to megavolt conditions, electrical discharges are sprayed into the atmosphere—toward an electrical ground if one is available, or otherwise simply out into space.

If a person holds a large conductive object such as a bar of metal in the presence of an activated coil, Tesla said, the holder feels nothing but perhaps a slight resonant buzz. (Tesla did this himself for the fun of it.) Despite the extreme voltages, the high frequencies, said Tesla, were unable to penetrate skin.

Tesla was able to bring evacuated globes and electrical tubes to glowing brightness simply by touching them when he was electrified by the field of his apparatus.

good circumstances, the sparks it generates leap around 25 feet. Physically, it is a big machine—16 feet high and 4,800 pounds. It takes a crew of three men the better part of two days to set it up.

Jeff's other favorite coil, the slightly smaller SG-20, is unique in that it was designed and built to hang upside down from the arena ceiling at NHL hockey games. As originally envisioned, it would hover over the penalty box of the Colorado Avalanche hockey team, adding a little fire over the ice to engage the crowd during the pregame warmup. Just before the home team would hit the ice, the lights in the Pepsi Center would dim dramatically and spotlights would sweep the stands. An artificial mountain would move onto the ice with the big SG-20 suspended above it, the capacitor bank charging and the motors that control the frequency of the spark timing whirring. After an appropriate pause to build tension, the coil would crackle to life. Lightning bolts and streamers would flicker across the space above the darkened rink, a simulated electrical storm bathing the stands in reflected electric fire.

Unfortunately, the coil was a bit too extreme for NHL officials. The noise and lightning made people nervous, most of all the Pepsi Center's multimillion-dollar scoreboard manufacturer, whose engineers feared what the big coil might do to the circuitry and components inside the scoreboard hanging over center ice. After a few months of corporate pressure, the inverted coil was

---

According to the account of writer F. Scott Taylor, "The individual bolts of hissing electricity shooting from the points on his body might be three feet long or longer, and as thick as a boa constrictor. Using a metal plate, Tesla would cut the severed lengths into fireballs. Then he would toss the sizzling remains to his spectators."

Similarly, wearing boots with five-inch cork heels and a formal tuxedo complete with white tie and tails, the original Technology Underground entertainer would control the flow of electricity through his body and bounce the electricity away with a piece of metal. He would stand on a platform elevated to potentials over 100,000 volts. There, sparks would shoot away from him, and if he held his hands a few inches apart, sparks would fly between them. At the highlight of his demonstration at the World's Fair of 1893 in Chicago, he passed 200,000 volts of AC through his hands, causing sparks to emanate from his head and hair in wild efflorescence. The crowd went wild.

THE TECHNOLOGY OF TESLA COILS

A coil is not a particularly easy device to explain, and even the most fundamental understanding requires a more than basic understanding of electricity. A very-high-level description of the way a coil works is this: The coiler (as the people who make these things are called) builds two electrical circuits close together. Circuit one—the primary—consists of a high-voltage AC source (such as a neon-sign transformer), a large and electron-commodious capacitor, a short inductor or coil made from thick copper tubing, and a spark gap. Circuit two—the secondary—consists of a very long coil of thin wire, with one end grounded and the other attached to an aluminum sphere or toroid placed on its top.

Ⓓ MOTOR TURNS DISK, OPENING AND CLOSING CIRCUIT

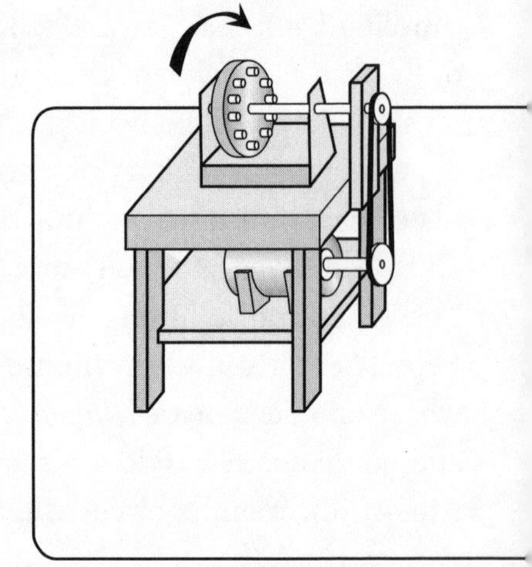

1. Tesla coils are air-core-resonant transformers. At resonant frequency, very high voltages are present at the electrode **(F)**, capable of ionizing the air to make arcs and sparks.

2. The transformer, capacitor, and coil (items **A, B,** and **C**) make up the primary network. Items **E** and **F** make up the secondary network.

3. The motor rotates a disk **(D)**, making and breaking the circuit at high speed. The interaction of the circuit's electrical characteristics with the spark gap sets up a ringing oscillation.

4. Energy radiates from the primary coil **(C)** to the secondary coil **(E)**. At resonance, energy in the secondary coil grows until it is great enough to produce high-voltage arcs.

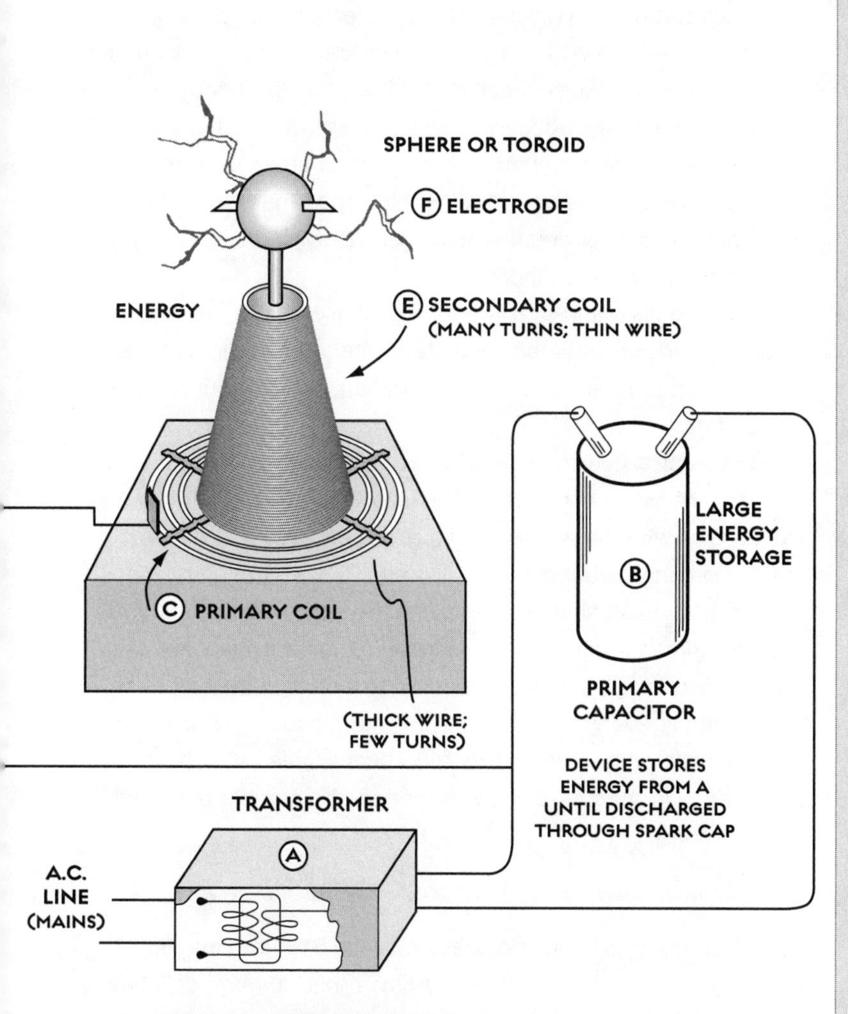
SPHERE OR TOROID
F ELECTRODE
ENERGY
E SECONDARY COIL
(MANY TURNS; THIN WIRE)
LARGE
ENERGY
STORAGE
B
C PRIMARY COIL
PRIMARY
CAPACITOR
(THICK WIRE;
FEW TURNS)
DEVICE STORES
ENERGY FROM A
UNTIL DISCHARGED
THROUGH SPARK CAP
TRANSFORMER
A
A.C.
LINE
(MAINS)
TRANSFORMER STEPS UP
LINE VOLTAGE FROM 120 VAC
TO AROUND 10,000 VOLTS

High-voltage AC power is applied to the primary circuit, and if the circuit is designed exactly right, the electrical current starts to oscillate back and forth in the primary coil and sparks start to jump across the spark gap. Each time the high-frequency spark appears, this has the effect of closing the circuit. When the spark disappears, the circuit opens. Thousands of times each minute, the white-hot spark leaps from one of the spark-gap electrodes to the other, effectively creating and then dissolving a conducting path through the atmosphere.

As the opening and closing, or "ringing oscillation," occurs, big electrical fields are radiated out of the primary coil and into the surrounding space, like a gigantic, enraged, out-of-control radio transmitter.

In close proximity to this madly oscillating high-voltage primary circuit is the secondary circuit, the very long coil of thin, closely wound wire with the aluminum doughnut set atop it. If the whole system is constructed correctly, the energy from the primary coil almost magically transfers directly to the energy-hungry secondary coil, and as it does so, the voltage in the secondary circuit is transformed, amplified, or stepped up to incredible levels.[1]

In normal, everyday transformers, like those found on top of utility poles, voltages are stepped up by a simple ratio: the number of turns in the primary coil divided by the number of turns in the secondary.

$$\mathbf{Voltage}_{\text{Primary}}/\mathbf{Turns}_{\text{Primary}} = \mathbf{Voltage}_{\text{Secondary}}/\mathbf{Turns}_{\text{Secondary}}$$

For example, if 1 volt appears across the 5 turns in the primary coil, and the secondary coil has 3 times that number, or fifteen turns, then the voltage across the secondary is 3 volts.

The interesting thing about a resonant air-core transformer, as Tesla figured out, is that the voltage increase in this device is determined by a much more complicated equation, and that's because the Tesla coil achieves a great gain in voltage in a very different way than a conventional transformer. Instead of a simple ratio based on the turns, or windings, in a transformer, a Tesla coil's voltage gain is based upon the different impedances of the primary and secondary circuit components.* The math is more complicated and involves invoking the conservation-of-energy principle, but a rigorous analysis shows the voltage gain is given by this equation:

$$Vs = Vp\ (Cp/Cs)^{1/2}$$

where Vs and Vp are the secondary and primary voltages and Cs and Cp are secondary and primary capacitances, respectively.†

In a Tesla coil, the system voltage gain increase is determined by this mathematical equation—one that involves exponents and the ratio of capacitances, resulting in extremely high multipliers. This is another example of Tesla's genius, for in a resonant, well-tuned coil, the voltage potential at the toroid can be in the range of a million volts.

Millions of volts in electrical potential may be spread between the aluminum toroid at the top of the coil and the other end of the

---

*Impedance is the alternating-current world's analogue to the direct-current concept of resistance. That is, it's a way to represent how much current will flow through a circuit, given a specified AC voltage across a circuit load. If you have 1 volt of AC across an impedance load that lets 1 ampere of AC current flow, the impedance is defined by the AC version of Ohm's law and is 1 ohm.

So why use impedance instead of resistance? Because alternating current not only has amplitude, like DC, but also has frequency and phase characteristics. Impedance takes into account that the load will not only allow a current to flow, but will change the current's phase and frequency as well.

†Think of a capacitor as the electrical analogue to a water tank. The flow of electric current in and out of a capacitor works similarly to the flow of water in and out of a tank. A capacitor stores energy when a battery charges it, and a tank stores water when a pump fills it. The amount of electrical charge in the capacitor would be analogous to the volume of water in the tank. The height of the water in the tank would be analogous to the voltage applied to the capacitor, and the area of the tank would be similar to the capacitor's capacitance. A tall, skinny tank might contain the same amount of water as a shallow, flat tank, but the tall, skinny tank would hold it at a higher pressure. Similarly, a tall, skinny capacitor with high voltage and low capacitance would hold the same amount of charge as a shallow, flat capacitor, but hold it at lower voltage.

coil wire, firmly planted in the ground. The atmosphere breaks down into plasma in the face of such astounding voltages, and visible, audible, and odorous phenomena such as lightning-bright arcs and streamers, 120-decibel static, and ozone are produced. The Tesla coil simultaneously assaults all the senses, and it is hard, very hard, to spend much time close to a working coil without sensory protection. Most important of all, the coiler must take great care not to involve the sense of touch, because with the megavolt arcs just a little touch may result in the coiler becoming just a little dead. <

removed, given hockey's equivalent of a Pete Rose–style ban for life. Like Pete Rose, the SG-20 continues to show up occasionally on television and movies in a variety of roles and places.

It's hardly surprising that people found the huge hockey-arena coil a little intimidating. There is absolutely no doubt that a big, operating Tesla coil is a scary thing. Jeff powers up the SG-20. It is loud, it is bright, and the exact location and length of those bolts of highly excited ionic particles are unpredictable.

"The bolts seem to reach out toward you," yells Parisse. "It's like they're alive and have a mind of their own. The bolts look like they want to reach out and grab you."

> > >

You climb aboard the Monsters of Voltage tour bus for the 1,500-mile drive to the east. Your destination is a modest home sitting on five acres in the quiet town of Newcastle, just south of Oklahoma City. It's squarely in the middle of Tornado Alley, the place where tornadoes spin up out of unstable air masses.

Tornado Alley is the midwestern version of the San Andreas Fault—a place where tectonic-plate-like collisions of large air masses occur over the ground instead of underneath it. In spring, the airborne earthquakes—the big thunderstorms—spin up with regular frequency. The skies often fill with roiling thunderheads throwing off cloud-to-ground lightning. Your host, Kevin Eldridge, grew up in Tornado Alley; storms, lightning, and thunder have always been a part of his life.

In May 1999, to the wail of the civil defense sirens, Eldridge gathered his family and took off for the hills, to avoid "the big one"—an F5 tornado, a mile-wide monster that cut through the heart of Oklahoma. Eldridge got a good look at the enormous black cloud hugging the ground, the single most destructive tornado, in terms of dollars, in history. In fact, some University of Oklahoma storm researchers say they have evidence that the 318-mph wind speeds clocked by sensors placed in the cyclone's path might qualify that tornado as the only true F6 tornado on record. F6 or not, the power of the big one remains legendary, even in Tornado Alley.

But luckily, you've picked a quieter night for a visit, at least in terms of weather. As you unload your belongings from the tour bus, the evening air smells fresh, slightly moist, and cool. This is a place far, far away from the lights, smog, and traffic of California. In the backyard of a modest frame home stands a nondescript metal shed. And inside the shed is Kevin Eldridge's BIGGG coil, the fourth largest (by some measures) Tesla coil in the entire world. It is 16 feet high and capable of producing 26-foot-long arcs.

It is mounted on a movable cart, so whenever Eldridge wants to fire it up, he merely needs to open the shed's oversized door, push the cart out into the yard, connect up a few cables, and turn it on. It's a swift and easy ten-minute procedure from start to finish. In a great stroke of coiler's luck, there's even an Oklahoma Gas and Electric high-tension line running through the Eldridge property, and from there he can obtain the 200 amps of current at 480 volts he needs to throw 26-foot-lightning bolts.

Eldridge's is likely the largest coil that is strictly amateur—that is, it never has and probably never will produce any income for its owner. It is a true hobbyist coil, designed from scratch with parts scrounged from wherever they could be found the cheapest—going-out-of-business sales, auctions, and the Internet. This is a machine nearly as tall as a farm silo, capable of generating two-and-a-half-story-tall arcs. It's a custom-made 55-kW power sink out on the prairie, and it cost less than $1,000 to make.

The typical large hobbyist-sized Tesla coil consumes somewhere between 1,000 and 10,000 watts when operating. The BIGGG coil uses more than 30,000 watts, and converts that energy into arcs and sparks that sometimes, depending on weather conditions and how hard Eldridge wants to push it, exceed 26 feet. His neighbors used to call the fire department, worried that there was some type of big electrical fire taking place on his property, but that doesn't happen anymore.

This is a good night for a Tesla coil display. Eldridge powers it up and undoes the safety interlocks. The electrical potential from toroid to ground builds and intensifies. Nearby lamps fluoresce, hair stands on end, and then the sparks begin to fly. White-blue cascades of sparks and arcs froth down from the toroid to the ground and into the air. Some watchers consciously stifle their impulse to run and hide. The arcs increase in size. They extend up and around the shed and around his backyard, lighting up the sky with Tornado Alley–style energy.

> > >

The tour bus pulls up and parks outside Bill Wysock's suburban southern California home. From the outside, it looks unremarkable. But the inside is a museum, full of interesting electronic gizmos: vacuum tubes, frictional spark generators, and enough 1960s electronic equipment to make the room resemble a NASA Mercury space shot control room.

But it's Wysock's Tesla coils that are the focus of this tour, and he's built many. Wysock's coils are serious, state-of-the-art, powerful, and highly optimized. His biggest, the one he refers to as 13M, is currently stored in carefully packed parts. But it's not at his house, because the neighbors aren't too fond of it. A few years ago, Wysock says, "I was shooting 35-foot arcs over the backyard. You could see them for two or three blocks. One neighbor came over and said, 'You and your bleeping lightning bolts scared my twelve-year-old so much he urinated in his pants.' " So the tour makes a side trip to a small airport, where the 13M is stashed among the Cessnas and Beechcrafts in an aircraft hangar. The toroidal top load (the doughnut-shaped electrode from which the lightning appears to leap) and the primary and secondary inductors are crated and ready to go, just waiting for the next opportunity to be erected. It is not often assembled simply because of the work and expense involved in setting it up, which requires a crane, a forklift, a 40-foot semitrailer, a bobtail truck, a crew of five, and three days.

But when it is set up for exhibition and events, it looks a lot like a Buck Rogers–style water tower leaking huge electrical torrents. The 13M cracks out 50-plus-foot arcs with ease.

The final leg of the trip follows I-5 to San Francisco, birthplace of the Electrum. Engineer Greg Leyh's Electrum is a four-story-high coil/artwork. When the switch is closed, a whopping 130 kilowatts of power (more than any other coil, and four times the power consumption of Kevin Eldridge's back in Oklahoma) blasts through the coil's circuitry. Unfortunately, the tour group can't get a look at Electrum. In fact, few people, even those heavily involved in the international Tesla coil community, have actually seen it, for it lives quietly on a wealthy art patron's sheep farm in New Zealand. The farm belongs to one of the country's richest businessmen, and the coil was commissioned and built as a piece of modern kinetic art. Electrum's specification sheet says it draws slightly more power than 13M, although the arc length claimed is just slightly shorter.

The coil's owner is out of the country frequently, and often for months at a time; therefore, it is the farm's superintendent who turns it on for testing and maintenance at intervals. The towering coil stands in a secluded spot near the coast overlooking the South Pacific. It is said that once a month, the caretaker fires it up and shunts an aluminum smelting plant's worth of juice directly into the tank circuit, which is about the equivalent of turning on all the lightbulbs in a whole neighborhood simultaneously. Electrum whirrs, cracks, and pops, firing lightning bolts and ionizing the surrounding atmosphere into long, steaming rivers of blue and purple electric arcs.

Greg Leyh is a very ambitious engineer, even by coiler standards. He has built several unnamed coils of extreme size and power, which are displayed in museums and operated occasionally in exhibitions. Although his Electrum is an incredibly large coil—by some measures the world's largest—he has bigger plans. His goal is to raise several million dollars to design and build the ALF, or Advanced Lightning Facility, which is basically a set of two monstrously big, ultra-high-voltage Tesla coils. How big? The ALF is designed to blast 5 million watts through twin 150-foot-high, oppositely charged Tesla coil towers.

According to Leyh, such a facility would be immensely useful in a variety of ways. The experiments it would enable would add to the store of basic knowledge about high-voltage electronics. It would be useful for testing the lightning resistance of airplanes, rockets, power grids, and so on. But its real value in the Underground is simply technology as entertainment, a testament to scientific fun.

The paper design for this coil portrays a pair of fifteen-story-high towers separated by 300 feet of high-voltage no-man's-land. The towers consist of miles of tightly wound copper wires. The pedestals house supersized banks of capacitors. According to Leyh, the design for the coil represents the maximum conceptual size for any coil, real or theoretical, because of a size-limiting factor called the "square-cube law."

Galileo was the first scientist to write about the square-cube law. At some point during his investigation into the fundamentals of motion and kinematics, he made an important detour into a brand-new engineering region, namely, the study of a branch of science called "strength of materials." If a structure increased in all dimensions equally, he said, it would hit a natural limit: any bigger and it would collapse due to its own weight. This is true for buildings, machines, even people. It is also true of Tesla coils.

Leyh thinks that his coil design for the ALF describes a coil that reaches the upper limit in both size and power. For Tesla coils, built from spun aluminum spheres, synchronous AC motors, and copper wire, Leyh's calculations show that 5 million watts of power and a streamer length of roughly 150 feet is the upper size limit. According to his figures, it can get no bigger.

On a paper napkin, Greg roughly sketches out his idea for the two big electrical towers that form the ALF. According to these figures, Greg's design can handle 5 megawatts of raw AC power. That's an awful lot of power, the equivalent of the average electrical power demand of, say, Grand Island, Nebraska, or White Plains, New York. And this is an idea for a single machine that with the flip of a switch swallows all that and more. The ALF coils, says Greg, would represent the largest coil that can be built, by his calculation, on this planet.

"There is a sense of permanence to this idea because once I do it, it's done, all done," he says. "Because of how a coil is constructed, and the limitations of the materials in it, no one could ever build a bigger one. It is a record that will stand forever. Period."

4. high-voltage DISCHARGE MACHINES

The slight twinge in your back is a reminder that you've spent the greater part of yesterday digging a large hole in the hard, dry earth of Brian Basura's backyard—spent the afternoon burying radiators and old metal parts in shallow graves. The heavy and electrically conductive metal parts were wired together with thick copper cable for intimate electrical contact. Then you tossed them down there, not so gently, and for good measure watered the earth with as much moisture as could be obtained from a half-inch-diameter garden hose on full blast. But the soil here in southern California is exceptionally dry and sandy and it appears that the capacity of the soil to absorb and dissipate water is far greater than your ability to make it noticeably wet. After a while, it does get kind of damp and muddy, and you hope that will be good enough.

The soil of southern California is good for a lot of things, such as, say, growing vegetables or mining alkaline minerals. But it's not good for obtaining an effective electrical ground. And today is the southern California Teslathon, and the one thing you and your fellow high-voltage enthusiasts want more than anything else is a dead-solid electrical ground.

Teslathon is the name given to the rallies attended by members of the high-voltage hobbyist community. There are Teslathons held on ad hoc schedules around the world—Texas, Wisconsin, New York, Virginia, England—wherever high-voltage enthusiasts live, for most of them relish an opportunity to show off their machines and ability and compare notes. This gathering is among the largest; there are dozens of machines here. The organizers have spent days preparing the physical space, the most important requirement being adequate power and electrical grounding.

Despite the name, a Teslathon isn't just Tesla coils. It is a more generalized gathering of electrical hobbyists who specialize in making and operating high-voltage electrical apparatus, generally with the purpose of obtaining long, noisy electrical arcs.

There are many different methods and machines that can accomplish this feat besides the Tesla coils. The festival features a great variety of devices, the common thread being their use of very-high-voltage electricity and their entertainment value. A Marx generator, for example, is basically a cascading series of capacitors and spark gaps, laid out in such a way as to produce a voltage-multiplying effect resulting in a single, massive, powerful, destructive spark. There are other high-voltage arc-making machines as well—Van de Graaff generators, Jacob's ladders, magneformers, and a few other less-well-known devices.

You've come to this big backyard in the horse-country hills north of the LA basin to mix with other enthusasists and have a day of arcing and sparking. During the daylight hours, the high-voltage experts come and set up their machines, each taking care to obtain good, reliable connections to power and to ground. Then, as evening falls and the darkness gathers, it's time for them to strut their stuff. There are nearly thirty machines here, some quite large. The most anticipated coil is the biggest, Jeff Parisse's SG-20, a two-story Tesla coil capable of shooting 20-to-30-foot arcs.

The smaller machines run first. The Marx generator wakes up the neighborhood with its sharp report. The Van de Graaff generator sends an overflow of electricity cascading from its mushroom-shaped head. The other Tesla coils do their thing. One after another they buzz and spark, producing long fingers of violet-tinged fire. And now it's time for the big one, the SG-20.

The SG-20's control panel glows in the moonlight, its bank of indicator lights showing the current status of each component. Following a rigid checklist, developed through years of high-voltage experience, Parisse releases the safety interlocks and turns the knob to start the flow of electricity into the primary transformers. The machine starts to hum. Small trickles of electricity begin to dance on the top electrode. They lengthen and stream outward, up toward the sky and out into the air.

Then something unexpected occurs. Above the din made by the electrical field of the big coil, there is an explosion, a loud one, and then the sound of many things breaking. The indicator lights don't show any problem whatsoever, but that's only because they're gone, blown to bits by an electrical surge going the wrong way in this circuit.

"Wow! That didn't feel good," says Parisse. "Is everybody okay?"

> > >

Even when there is no Teslathon taking place, Brian Basura's garage is a pretty interesting place. It holds an extensive collection of scientific equipment that is vaguely threatening-looking due to the many orange-and-black warning decals with the words DANGER—HIGH VOLTAGE stenciled upon them. The Basura backyard has been the scene of many interesting happenings, ranging from electrostatic lifter flights to Teslathons. But one of the most unusual experiences is today's demonstration—the operation of a particularly clever machine called a quarter shrinker.

Demonstrating the action of the quarter shrinker requires a lot of voltage, an elaborate safety procedure, and some spare change. As Basura makes the final check of wires and connections, his audience heads into the garage for cover. Inside, they consider the thickness of the walls and the sturdiness of their construction. One person wonders aloud, "Is that wall strong enough to fend off arc-flash-induced shrapnel?"

Basura stands in the garage door opening and motions that now is the time.

"Okay," says Brian, "here's where we need to be careful." Brian stands just inside the door jamb of his garage, barely beyond the possible trajectory of shrapnel that could emanate from a malfunction of the boxy electrical contraption humming in his driveway.

The focus of his attention is an assemblage of gray and silver high-voltage electrical parts mounted on an L-shaped frame made from perforated and expanded steel. The thing that sets this particular piece of gear apart from the other homemade high-voltage stuff, such as Tesla coils, is the heaviness of its construction. Most parts of it are crafted from steel diamond-stamped plate, a quarter inch thick, and the wire connections from one part to another are broad, ropy copper wires. Attached to the frame is a bank of capacitors, large enough to float a beefy 20 kilojoules of energy between the dielectric plates swimming in oil inside. Indeed, this thing looks like it could fire up a bank of arc welders.

This is no ordinary experimenter's electrical assembly—no fragile and delicately rendered computer circuit boards to be found here. This is Brian's invention—an electrostatic discharge machine built to do a specialized job. It rearranges the internal metallic structure of metal objects. It dumps city-sized loads of juice, current, amperage, electron flow—however one cares to consider it—through the internal crystalline structure of an engraved, disk-shaped blank of copper-nickel alloy. This piece of equipment shrinks U.S. legal tender—Washington quarters—literally, instantly, and powerfully.

When the quarter shrinker is activated, it throws huge amounts into the coin in a very short time interval. This, in turn, induces huge currents in the coin. As the great English scientist Michael Faraday determined, there is a relationship between magnetism

THE TECHNOLOGY OF QUARTER SHRINKING

The quarter shrinker invokes a principle sometimes used in high-tech commercial metalworking but taken to a more extreme and entertaining degree. It does something technically known as "high-velocity electromagnetic metal forming," or what Brian Basura more succinctly calls "magneforming." The magneforming process deforms thick hunks of electrically conductive metal by squeezing it using rapidly changing, hyperpowerful, albeit very short-lived, magnetic fields.

The machine in Basura's garage consists of an array of fully charged capacitors, connected to a homemade but highly sophisticated current-controlling apparatus. The core mission of all the apparatus, electronics, safety equipment, and controls is actually quite simple: The machine takes a very large amount of electrical energy, stored in the capacitor bank, and shunts it through a high-capacity, ultra-fast-acting switch through a copper coil holding a coin.

Brian's machine is designed and built to induce extremely large amperages into the outer rim of the conductive coin. In fact, the current that's induced is estimated to reach a million amperes or possibly more. Because voltages involved are very high as well, for the fraction of a microsecond during which the capacitors dump their power into the coin, the instantaneous power coursing through the coin is off the chart. Power is defined as the time rate of doing work, mathematically expressed as

Power = Energy/Time

If a process uses 6 joules of energy in two seconds, then the power involved in this process is simply 3 joules per second, which is the same as 3 watts. If the same process used up the same 6 joules, but now in just one second, the power would be measured at 6 watts. If the energy is used up in shorter and shorter intervals, than the instantaneous power consumption, the wattage, goes up and up. <

and electricity, and few machines illustrate this relationship as powerfully as a quarter shrinker.

Here, the electricity surge induces powerful magnetic fields that go to work on the coin's exterior. In an instant, repelling magnetic fields between the coin and its holder create tremendous radial compressive forces; forces so large that they crush the atoms together, permanently deforming the quarter. The metal shrinks almost instantly to a smaller diameter.

Basura uses very large capacitors in his quarter shrinker. The plates in such a large energy storage device hold a lot of electrical energy, which is set loose as the capacitors are fully discharged in a fraction of a microsecond. This results in gigantic instantaneous power levels. Controlling that power is the job of a switch, and perhaps the most unique thing about Brian's machine is the switch he's designed. It's a round, machined hunk of metal he calls the "trigatron," an ultra-fast-acting switch that can shunt the electrical energy from the storage capacitors into something else in a minuscule amount of time.* According to Basura's calculations, for a

---

*Ultra-fast-acting switches are great for a lot of Technology Underground projects: making wires explode, vaporizing small metal objects, shrinking quarters, making illicit nuclear weapons, and so forth. The ability to direct a Niagara-like surge of electrons where, and more importantly, exactly when necessary allows the technological self-expressive to engage in all sorts of high-energy fun.

Everyone is fundamentally familiar with the basic operation of an electric switch. In the Off position, there is no electron flow between components in an electrical circuit. In the On position, there is. Further, it is easy to understand that big switches with big chunks of conducting copper wires are able to handle higher voltages, greater amperages, and, simply, more electrons. A switch controls the flow of current in a circuit in a manner such that when the switch is in the On position, the current flows at a rate determined by the electrical needs of all the rest of the electronic components downstream. When Off, the current does not flow at all.

So, can something as fundamentally simple as an electrical doorway get complicated? Indeed it can. That's because the devil is in the details, in what happens in those fractions of a second during which the switch moves from Off to On. A switch has its own specifications and internal restrictions that determine how fast it can switch from open to closed, or how

very, very brief and shining moment, the amount of power that Brian's box is sending through that quarter is about equal to the amount of amperage supplied by a 10-megawatt coal-burning power plant.

As might be expected, letting loose such an unbridled flood of electrons within the outer circumference of the quarter has dramatic effects. The logical assumption would be that the quarter would simply melt or vaporize. But something very different occurs. The quarter experiences a staggering, atom-crushing magnetic compressive force. The coin is squashed, mashed, and compacted in a fraction of a second.

Inside the containment box, with ultra-high currents eddying around the circumference of the coin at light speed, the repulsion forces between the coiled work holder and the coin create a huge radial inward compressive force on the coin that easily overcomes the yield strength of the metal. This causes it to plastically deform, and the coin morphs to a much smaller diameter and thicker cross-section. At the same time, a similar but outward radial force literally explodes the work coil with the energy of a small bomb.

Once the coil disintegrates, any residual magnetic or electrostatic

---

rapidly it can interrupt the flow of current once it has been opened. Ideally, the current goes from nothing to full-on instantaneously.

The mechanical switches such as light switches that are found in homes exhibit no such ideal behavior. The time taken to switch from Off to On is considerable, measured typically in the millisecond range. In addition, there are other unintended effects going on, such as bouncing (which occurs as the switch fluctuates rapidly from open to closed in the process of being physically manipulated by the operator's finger). The relatively long time it takes to go from Off to On and the bouncing doesn't matter when a person turns on a washing machine or flips a light switch. But to the people who make quarter shrinkers, wire exploders, and nuclear weapons, it absolutely does matter.

There is a school of contemporary corporate science that professes the obsolescence of mechanical switches and vacuum tubes. The typical line of reasoning is that the advent of solid-state electronics should have turned the issues regarding high-current, fast-acting switching into a relic of the past. This is not completely true because semiconductors are limited in other ways, particularly in that it is very hard to manufacture semiconductors capable of switching kiloampere-sized currents.

energy remaining in the system is transferred into a swirl of hot blue-white plasma that bangs against the steel-plated walls of the containment device.

It is time to see what this box can do, so Brian places a new, shiny quarter in the containment box and steps back. Way back. Last year, something went wrong and the machine blew up there on his driveway. So now he's extra careful. He pauses and then takes a minute to preface the show by describing what happened last January.

"I was in the driveway shrinking some quarters for fun and really had everything working the way it was supposed to. All my quarters were turning out perfectly, every one of them. I suppose I got a little complacent, maybe a little bored. So I decided to turn it up, to see what would happen when I really turned on the power.

"I started to charge the cap for the shot. Luckily, I had a moment of clear thinking and put on my full face shield and tossed my flak jacket over my winter coat. The plan was to run it up to about 26 kilojoules."

Brian pointed to the top of the bank of capacitors, as if they were tanks for storing gasoline. "Oh yeah, that's a lot. A whole lot.

Therefore, there are still instances where a person needs to go beyond what is provided by modern electronics hardware designers. To make a device to shrink quarters, thousands of amperes of current must be sent in nanosecond activation times at electrical potentials of thousands of volts. That can't be done with anything less than the fastest, most rugged electrical switches possible, and surprisingly, that often requires going a step back in time, to the pre-silicon, pre-solid-state era, to the days of the vacuum tube. It requires a switch that makes use of an electrical apparatus called a "triggered vacuum spark gap."

The triggered vacuum spark gap is a simple concept: A high-voltage pulse called the trigger pulse is applied to an electrode. This in turn initiates an arc between two switch terminals called the anode and cathode. When the trigger pulse goes off, the arc is struck, and electrons—a lot of electrons—travel through the ionized arc all at once. This is the same phenomenon as the arcing in the spark gap of a Tesla coil—the ionization of the air allows current to flow for the duration of the arc, and then stop.

A trigger pulse is utilized within the trigatron to initiate the main discharge, hence the name. Different types of homemade and commercial electronic switches have been

Things started out fine and the cap was charging like usual, but something wasn't right.

"The last thing I remember was seeing the needle move past 12.5 kilovolts and the power-supply upper-limit light come on. Everything went quiet, and *ka-pow!* The shock wave hit me like a truck.

"I don't think I was knocked unconscious, but I was disoriented and realized something had gone really wrong. I wasn't thinking too clearly. What I think happened was that one of the electrical contactors failed and the current flashed over the top and blew up some wires and other stuff. Luckily I only had a sore chest and neck, and my hearing came back after a few minutes."

And so from that time forward, Brian has been even more careful. After checking all the safety equipment, latches, interlocks, and grounding devices, he pushes the fire button. No explosion, no thud, no smoke. There's a single loud, booming, brisant crack. "That's it," says Basura. "Let's go look."

The door on the containment compartment opens and he looks inside. There's what looks like a bunch of copper cornflakes, the remains of the copper-wire holding coil in which the quarter was mounted. And there, on the bottom of the tray, covered with

---

designed by those who make ultra-fast-acting switches to create the main anode-to-cathode conducting arc. They generally have interesting, scientific, and inscrutable "tronic"-sounding names, such as ignitrons, thyratrons, krytrons, and trigatrons.

In addition to the wholesome fun of quarter shrinking, high-power pulsed-voltage switches such as the trigatron also have a dark side. The detonators that fire high-explosive implosion systems in various types of nuclear weapons require this type of switch.

Consequently, commercial sale of and international trade in these devices are highly regulated. There were attempts to circumvent these regulations by spies and morally bankrupt arms traders in the 1980s and '90s involving one particular type of device, the krytron. So krytrons have become relatively famous, at least as far as cloak-and-dagger high-voltage electrical components go. For example, in Roman Polanski's 1988 movie *Frantic,* Harrison Ford's wife accidentally switches her suitcase with that of some international arms smugglers and winds up with a bagful of trouble in the form of a nuclear-grade krytron—a pulsed power switch that can shoot enough juice to plasmatize the metal foil that starts the atomic chain reaction in a nuclear weapon.

insulator ash and copper crumbs, is the coin, which has been completely transformed. The quarter has shrunk to the circumference of an undersized dime, but the cross section has thickened out, so now Washington looks thick and heavyset; it's as if we put a squat, labrose, tough-looking guy, like Al Capone or Yogi Berra, on our twenty-five-cent coin.

The quarter shrinker has rearranged the metallic structure throughout the quarter, cleaving through the crystalline structure of the metal and forcing it to bulge out along its Z axis. The coin has assumed a new and bizarre shape, yet it still maintains a slight resemblance to its old appearance as a U.S. quarter.

> > >

"Is everybody okay?" asks Parisse again, in the aftermath of the arc flash. Thankfully, no one is hurt. An arc flash is a sort of airborne short—an unwanted current flow making its own path. It is the high-voltage enthusiast's analogue to the high-power rocketeer's CATO. When insulation or isolation between electrified conductors is breached or can no longer withstand the applied voltage, an arc flash occurs.

At its worst, the temperature of an arc momentarily reaches more than 5,000°F as it creates a brilliant flash of light and a bang of tinnitus-producing sharpness. In large industrial applications, an arc flash can explode with an intense and time-compressed release of concentrated radiant energy, spewing hot gas and molten metal outward from electrical equipment. Worse yet, an arc flash creates a pressure wave that can damage hearing and can send loose material such as pieces of equipment, metal tools, and people flying. In high-voltage, high-amperage industrial accidents, which involve far higher energies than what is available here, a really bad arc flash can approach the destructive power of several sticks of dynamite.

But nothing as energetic as that has occurred here. In fact, after making a thorough check, it appears the extent of the damage is limited strictly to the blown-out indicator lights on the SG-20's control panel. The heavy-duty, hardened control components, transformers, and other items are undamaged.

The arc flash was due to the arid soil conditions in the area. Despite the connections to the various metal objects buried deep in the ground and irrigated with as much water as was available, the grounding simply wasn't good enough. A rogue electrical current jumped backward through the circuit, blasting the control panel.

The new control panel will not include indicator lights.

> > >

Members of the Technology Underground play with dangerously high voltages for reasons that may make them seem a bit mad to others. It's the idea of asserting control over stuff that seems almost uncontrollable.

Lightning is a wild animal. It does what it wants to do, where and when it wants to. Still, there are clever individuals who can tame lightning and turn it into predictable, if not fully controlled, high voltage, much as early humans turned wolves into cattle dogs.

High-voltage enthusiasts understand the personality and motivations of electrons. They know how to make them follow a copper pathway, queue up, and wait in orderly fashion, even jump across the air. It's an art more than a science, and if you really think about it, it's quite extraordinary.

5. hurling MACHINES

"Pie! Goddamn it. I had a feeling about that pumpkin," says Dean, your team leader, angrily. "Can't you pick a goddamn pumpkin that won't pie?"

It wasn't your best moment. When you pulled back on the release lever, the thing that you didn't want to happen did happen. The 9-pound pumpkin, round and seemingly firm when you loaded it into the sling, couldn't handle the centrifugal forces acting upon it as it was whirled around and around by the rotor. It probably turned to mush even before the trigger was released. You picked it out, so it's your fault. Lucky for you, there are still a couple more chances to redeem yourself.

It is a clear day and an unseasonably warm one for the beginning of November. You are standing in the middle of a recently harvested soybean field in the hinterlands of sparsely populated central Delaware, just outside of the town of Millsboro. The field encompasses several hundred acres, and all around the perimeter of the field there is a buzz of activity, a hum of people and machines hard at work making something big happen.

There are machines on the periphery of the field that rock back and forth, machines with arms that spring shut like big mousetraps, and machines with wooden arms that faintly resemble seesaws.

All are interesting, but what everyone is really enthralled by are the long arms of the 30-foot-tall, high-speed, windmill-like contraptions whirling around. These are the mighty centrifugal pumpkin throwers. They chug and puff, they whirr and spin. They are noisy, making a beating sound, like a damaged helicopter. Exciting and powerful, they're impossible to ignore. They command attention, and people are drawn to them. That is why you and the rest of your team decided to build one for yourselves.

These machines, the big centrifugal pumpkin hurlers, are echoes of the past, of a time before virtual reality, before computer games and CGI movie effects. These are tangible and authentic, set up methodically and solidly out there on the firing line. These constructions are set on thick wooden planks, steel I-beams, and anchors dug deep into the black earth, to contain the forces acting on the foundation when the machine operates. This piece of machinery is large, mechanical, and powerful-looking. It requires the use of a crane and a cherry picker to set up.

The centrifugal pumpkin thrower is a unique machine that is likely found nowhere else on earth but here. That's what is so interesting about the World Championship Punkin Chunkin Competition. There are several different and highly interesting types of machines in use here.

To the left of where you are standing, you can see the large oaken beams, rope hawsers, and heavy stone counterweights that belong to a group of machines classified as spring- and gravity-powered catapults. Massive and medieval, they appear brutish and crude, with their leather slings, rows of coil springs, and rough construction.

To your right are the air-cannon-class machines. They are big and powerful, yet the essence of simplicity. Mostly, they are big air tanks connected to big valves connected to long barrels, and not much more.

And in front of you are your machine and a few others like it—machines with names like Bad to the Bone and the De-Terminator. These machines resemble oil derricks or construction cranes that have reverted to their feral state. They are long, revolving arms mounted on spinning central pivots. They have large gasoline- and diesel-powered engines connected to the arms via weldments and bolts. An articulated framework of pivots, gears, belts, and drive shafts forms the transmission that transfers power from the engine to the spinning arm. Altogether, these components make up the machines that command the most attention at the event—the centrifugal pumpkin-throwing catapults of Delaware.

You and Dean and the rest of your team get three chances to toss an 8-to-10-pound pumpkin from the sling of the centrifugal catapult. The distance of each toss is recorded, and the team with the longest toss, in terms of horizontal distance, is the winner. Your first try is recorded in the official books as "pie," meaning that the pumpkin exploded into mushy fragments upon leaving the machine. This is the equivalent of a scratch in a track-and-field competition. But the good news is that there are still two more chances to make up for it, for only the single longest toss matters.

THE BEST HURLING MACHINES ON EARTH

The World Championship Punkin Chunk is a key stop on any tour of Technology Underground festivals. It brings together some of the most massive, powerful, noisy, and entertaining technology in the world. Over the next several days, the builders of cannons, catapults, and other hurling machines will compete for bragging rights.

In 1986, a somewhat eccentric and possibly drunk group of rural Delaware men challenged one another to a contest. They dared themselves to find out how far they could throw a pumpkin, using whatever means were available. As it turned out, the pumpkin didn't go too far. But Delawareans are a tenacious lot, and after a while, the small group became quite adept at pumpkin heaving. Something connected, and soon this group of friends organized a tongue-in-cheek, all-comers pumpkin-chunking competition.

At the initial event, the winning shot carried a 10-pound pumpkin for a ride of just less than 200 feet. Since then, things have gotten serious. Current participants go sleepless, foodless, and bathless in an all-out effort (or at least that's their excuse) to devise a machine to chunk their pumpkins farther and farther. And every year, they do go farther. The current crop of pumpkin-chunking artillery includes trebuchets, slingshots, onagers, spring engines, mangonels, torsion catapults, and colossal compressed-air-powered behemoths such as the 100-foot-long contraption called the Big 10-Inch.

Over twenty thousand spectators stand awestruck in a Delaware soybean field this afternoon to watch the engineer-creators of the Big 10-Inch and its eighty or so brother machines shoot, lob, and hurl pumpkins across Sussex County, all in the name of science, radical self-expression, and beer.

Sussex County makes up a substantial portion of the dangling lobe of mid-Atlantic peninsular land formed by parts of Virginia, Maryland, and Delaware. The locals call this area Delmarva, a contraction made from stringing together the first few letters of each state's name.

The salty tidewaters of the Delmarva Peninsula are inhabited by earthy, quotidian folk, raised on the ocean, bays, and marshlands of the area, where machines are freely and frequently modified to meet the needs of watermen and farmers. In the interior of the county, town dwellers work in food-processing plants, factories, and workshops, and they too are quite willing and able to wield drill bit and welding rod when the need arises. Out of a combination of isolation, boredom, unrequited creative instinct, and a civic mandate to attract more notice to an out-of-the-way part of the country grew an activity that became the second largest annual event in Delaware, surpassed only by the yearly NASCAR race at Dover Downs, 39 miles to the north.

How can an amateur pumpkin-chunking contest in a small Delaware backwater even play in the same league as mighty NASCAR, with its huge advertising budgets, saturation media coverage, and celebrity racecar drivers? It requires a visit to really understand the phenomenon.

The World Championship Punkin Chunk takes place over a three-day period, just after Halloween. The event is sited at a 400-acre harvested farm field outside the rural hamlet of Millsboro, about 50 miles inland from the Atlantic coast. It's a bit hard to find, since it requires taking back roads, and the event organizers don't bother with much event signage. But clever Punkin Chunk event-goers utilize their superior observational skills and watch out for a car or truck pulling a trailer full of 10-pound pumpkins or a 15-foot-high wooden catapult. Following such traffic will sure enough lead them right to the Punkin Chunk's main gate.

The grounds are divided into contestant and spectator areas. The contestant area bordering the field area, or the "firing line," as it's called, is a green curve of harvested soybean detritus that forms the edge of the projectile landing range. There is a buzz of frenetic building activity taking place on the line, with teams fine-tuning and making final adjustments to their machines. Downrange, out on the target field, there are no people, just splattered pumpkin carcasses: pale orange entrails and mounds of slimy pumpkin seeds.

The activity of pumpkin chunking may seem pointless to the uninitiated, but it is almost sacred to Technology Undergrounders who understand the allure of high-velocity vegetables in ballistic flight. The pumpkin chunkers and their machines are even beginning to draw attention from beyond the neighboring towns of Dover and Salisbury. On the days of competition, there are three different film crews present, representing three different television networks, in addition to a gang of print media reporters. This pastime seems to be catching on.

For scale, power, and gracefulness, the catapult is a true mechanical marvel: powerful, finely hewn, and primitive at the same time. Just by watching it work a couple of times, one readily sees why for more than eighteen hundred years, from about 400 B.C. to A.D. 1400, the catapult was likely the most complex, most expensive, and most important machine on earth. Wars were started and won, cultural boundaries defined, and the political ascendancy of nations determined all on the basis of who had the best catapults.

Ignored for about six hundred years, they've lately, well, catapulted back to fame. They are often the star attractions at festivals of medieval entertainment. They have been the special-effects focus of more than a few recent movies—from the *Lord of the Rings* trilogy and *Gladiator* to *Kingdom of Heaven* and more than twenty-five others. They show up frequently on television shows of all sorts. Ozzy Osbourne even used stage catapults to liven up his post–Black Sabbath solo concert tour by flinging glops of raw, glistening chicken liver into the cheap seats just before the encore.[1]

And there are a lot of people building catapults in today's Technology Underground. They have their own division at the World Championship Punkin Chunk. A group called the Society for Creative Anachronism has an active subgroup of catapult builders. And there are a host of unaffiliated catapult builders who construct their own trebuchets, onagers, and ballistae just for the pure joy of having one.

Despite the relative paucity of classical texts on technical subjects, the importance of catapults to ancient societies is visible in the number of age-old books and treatises on the art and science of catapult building. Ancient Greek writers such as Heron of Alexandria, Athenaeus, Apollodorus of Damascus, Polybius, Ctesbius, and Philon of Byzantium were the first to extol the power of catapults. Later, Appian, Josephus Flavius, and Ammianus Marcellinus made highly detailed accounts of the Roman Empire's catapults, and provided detailed directions for building them exactly as the Roman legions once did.

Of the Roman writers, perhaps the most prolific and important catapult documentarian was the renowned architect and master of Emperor Augustus' siege engines, Marcus Vitruvius. His most well-known work was a ten-volume history of all things related to Roman technology, including copious descriptions and advice taken from his own area of specialty—siege engines.*

Aside from all these Levantine sources, there are many more records and descriptions from all over the world—China, India, the Middle East—as well as northern Europe.

In spite of—or perhaps because of—the quantity of available information, the subject of catapult history is a little murky. The catapult was not invented by any single person or nation. Its design and application arose sort of coincidentally all over the world and no logical developmental timeline for the device exists. Most vexing of all is that scholars and writers at various times used different

*"Siege engines" is a general term that encompasses all of the machinery used in ancient and medieval times to besiege forts and strongholds. The equipment included not only hurling machines such as catapults, but weapons such as armored, movable towers from which attackers would jump down to do battle, and giant battering rams.

THE TECHNOLOGY OF ANCIENT CATAPULTS

There are four ways that ancient civilizations used catapults to send things flying: tension power, torsion power, traction, and gravity.

The earliest catapults built by the Greeks were much like very large bows and arrows. They shot short, heavy darts called bolts. They were powerful, but they were cumbersome and took a long time to load and fire. In general, these early catapults obtained shooting power from bending back a wooden or animal-horn bow or leaf spring. Such weapons were called tension catapults—it was the tension in the fibers of the bow that provided the motive force.

By the third century B.C., the Greeks started to experiment with different types of springs to shoot the bolts, and came up with the idea of using tightly strung coils of rope. This worked well, since the spring could be made very large and powerful by simply coiling the rope many times over.

A spring made this way gets its power by utilizing the energy locked inside that acts radially to its long axis, like a garage-door spring. This is called a torsion spring, and so catapults powered by coiled ropes are called torsion catapults. There are many Greek, Roman, European, and Middle Eastern names for torsion catapults: *chiroballista,* onager, scorpion, mangonel, and stone-thrower are a few.

The next change in catapult technology was the use of human power to shoot rocks. It may seem like a step backward to substitute human muscle for coiled springs and bows, but it really wasn't. The ancient engineers designed human-powered lever-based machines, known as traction machines, which were fast-operating and accurate. With a man or men pulling down hard on a lever, the machine could still hurl a rock about as far as the torsion and tension catapults, but now it had a firing rate several times faster than either.

Traction catapults were usually just called by the generic term "traction catapult," but other terms were sometimes mentioned in the ancient literature: pull-thrower, *hseng peng* (by the Chinese), or witches' hair (in the Near East).

Traction machines showed up in Europe somewhere around A.D. 1000 by way of China and the Middle East. Initially they were made out of a stout swinging beam that pivoted horizontally, with a sling at the shooting end and ropes at the other that were pulled simultaneously by a gang of men. When a man, or two or three men, pulled on the rope simultaneously, they could toss a rock a very long way. If ten or twenty men all pulled simultaneously, they could heave boulder-sized ammo.

Once introduced into medieval warfare, the traction machines really became the catapult of choice. Whereas the ropes and bows of the torsion and tension machines were affected by weather extremes—the ropes and bowstrings stretched and shortened, making it hard to shoot accurately—traction catapults worked no matter what the weather. It was an artillery piece for all seasons. A traction-style catapult could be loaded and fired at a much faster rate than spring-type catapults. For a torsion-powered catapult, the operators had to load a rock into the sling, cock the crossarms with a big winch, lock the arms in firing position, and then pull the trigger, all of which took a lot of valuable time. Conversely, with the traction catapult, it was just a matter of loading it up with a rock and pulling on the ropes. All that was necessary was a simple load-and-pull technique.

For these two reasons—all-weather capability and speed of fire—the traction catapults speedily replaced the old-fashioned torsion types. They were the best bang for the buck all over Europe, used with tremendous success in various sieges and crusades until about the year 1200.

The last type of catapult to be invented was the gravity-powered version. These machines used very heavy weights to flip a lever arm to which a rock was attached. When the weight bundle fell, it tossed the ammunition resting on the other end of the seesaw-like lever toward the enemy castle in a high, soaring arc. These machines were called trebuchets in general, but had other names, such as *couillard, petrary, bricole,* and *blida.* The French engineers named their gravity-powered catapult "*les testicules*" because it had two large counterweights hanging down from the long lever arm.

The gravity-powered catapults were the biggest of all catapults and, just like battleships, were often given their own unique names, such as the Bull Slinger, Tout le Monde, the Wild Cat, the Parson, and the Evil Neighbor.

Around 1200, a now unknown Arabian or European engineer asked himself this question: "Instead of having a host of big, strong men pulling on ropes, why not simply attach a great counterweight to the end of the lever?" If the master of engines did this, then he needed fewer men to work the device. And to throw a bigger rock, all he needed to do was increase the size of the counterweight. Even better, such machines were very, very accurate since the motive power was always the same for any particular weight. With men pulling, forces were harder to regulate and a projectile could fly farther or nearer than the master wanted.

Now, because the counterweight made the firing of a trebuchet so repeatable and so accurate, a well-commanded cata-

1. The throwing arm **(A)** is pulled down toward the ground and latched **(B)**.

2. The operator pulls on the trigger pull **(C)**, releasing the counterweight **(D)**.

3. The arm spins free and rotates. As it does so, it whips the sling around **(E)**.

4. Based on the machine's geometry, the projectile leaves the sling at the optimum position. One end of the sling flies off the notch **(F)**.

5. The projectile soars downrange **(G)**.

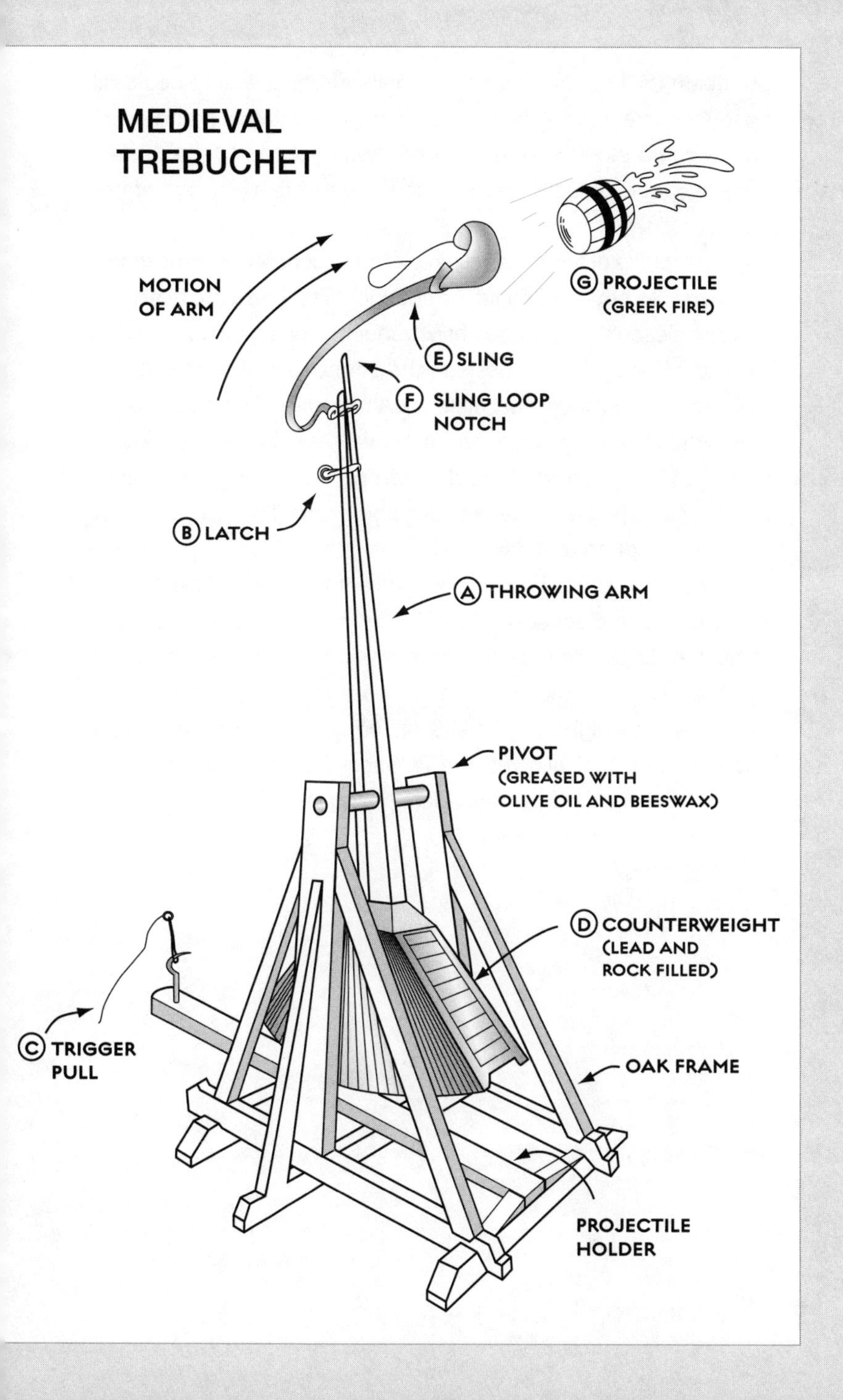
MEDIEVAL TREBUCHET
MOTION OF ARM
G PROJECTILE (GREEK FIRE)
E SLING
F SLING LOOP NOTCH
B LATCH
A THROWING ARM
PIVOT (GREASED WITH OLIVE OIL AND BEESWAX)
D COUNTERWEIGHT (LEAD AND ROCK FILLED)
C TRIGGER PULL
OAK FRAME
PROJECTILE HOLDER

pult team using good, hard stone balls of equal weight could hit nearly the exact same spot on a castle wall with each and every shot. Even a stone wall, made from hunks of sandstone 12 feet thick, would eventually crumble under a constant barrage of trebuchet stones.

The rapidity of firing was always a major concern for army commanders. So how fast could a trebuchet fire? There are historical records describing an immense trebuchet designed by Bishop Durand of Albi, used to besiege the stronghold of Montsegur in Italy during a religious war called the Albigensian Crusade in 1244. The fortress of Montsegur had rock walls several feet thick and was sited high atop a solid rock outcrop.[2] It enjoyed a commanding defensive position over any besieging army that was brave (or foolish) enough to try to attack it.

But even with all these natural advantages, the castle still fell to the Crusaders because of the power and accuracy of their big trebuchet. Durand's rock tosser threw rocks, some weighing as much as 175 pounds, at the castle walls at twenty-minute intervals, day and night, for weeks on end. Finally, the onslaught opened holes that allowed the fanatical crusaders to enter the fortress and doomed the defenders hiding inside. <

names to refer to the same types of catapults. The machine that the Greeks might term a *litho-ballista* the Romans might call a *tormentum;* the one that the English might term a trebuchet, a Frenchman might describe as a mangonel.[3] And even in the same country, people gave different names to catapults. So of the few catapult-related certainties, one is that the same basic throwing machine might have different names depending upon who is describing it.

The other thing that is certain is that there were many types of catapults. In general, any mechanical device built for the purpose of hurling missiles is known by the general name *catapult*. But this is a very broad term, so those who study them find it convenient to devise a taxonomy according to how they are powered.

PROJECTILES

One of the reasons catapults were such effective weapons is that they were extremely versatile in terms of what they could shoot. While large rocks were certainly a favorite projectile for demolishing castle walls, catapults could be designed with many different sorts of baskets, slings, and pouches. A siege engine could be designed to hurl just about any type of projectile that the engine commander chose.

In fact, records show that almost every heavy, hard, unpleasant, or disgusting object imaginable was used in catapult attacks: rocks, iron arrows, baskets of venomous snakes, diseased horse carcasses, clay pots filled with asphyxiating gas, flaming barrels of gooey chemicals, hornets' nests, the dead bodies of captured enemy soldiers, the severed heads of messengers, and cattle manure.

There was almost no limit to the creativity of ancient fighters when it came to selecting ammunition. There are historical accounts of two thousand full cartloads of animal manure being thrown at the siege of Carolstein by Lithuanian attackers in the later Middle Ages. Edward I of England was fond of using lead balls, which he made by removing the lead from the roofs of area churches.

If the battle plan was to knock down castle walls, then large heavy stones were preferred. The stones would be shaped into regular round balls of approximately equal size and weight so that they could be aimed reliably. But if the artillery was meant to attack wooden buildings or palisades, then they would often hurl barrels made of thin wooden slats filled with a burning liquid. The barrels would break against the fence and set it aflame.

As far back as 600 B.C., incendiary mixtures were used in warfare. Various concoctions were mixed together and hurled over the walls and gates of enemy fortifications. The ingredients in such mixtures included sulfur, pitch, naturally occurring petroleum, sawdust, and oils of various types and weights. The formulation of incendiary compounds was somewhat akin to making homemade chili—everyone had his own secret recipe. Some ingredients made it more sticky, some caused it to burn more intensely, and others made the fire harder to extinguish. Sulfur gave the mixture a horrible smell. Other ingredients, such as rock salts, made the fire glow bright orange. Medieval recipes for incendiary compositions included such unusual elements as oil of benedict (made by soaking bricks in olive oil), saracolle (a tree resin collected only in Ethiopia), and verjuice (the acid juice of crab apples, expressed and formed into a liquor).

Of all the fiery substances shot from catapults, the one used by the Byzantines was best known. It was called "Greek Fire" because it was first used by the Byzantine Greek armies. The ingredients are no longer known with certainty but likely included sulfur, tree resins such as tar and pitch, asphalt, petroleum, vegetable oil, turpentine, and powdered lime.

Grigg Mullen Jr. knows a thing or two about what is good and not so good to use as catapult ammunition. He's hurled aluminum beer barrels, 90-pound rocks, and oversized water balloons. Mullen-crafted catapults have fired softballs, milk jugs filled with water, water jugs filled with beer, wooden posts, bags of ignited tow, and all sorts of miscellaneous household goods.

It is quite possible that Mullen has built more large trebuchets than any other living person. Some of his catapults are parked in the back of his 20-acre farm in Virginia's Shenandoah Mountains. Others are in warehouses, in museums, or standing guard outside English castles, just in case the French or Scots start to get unruly and make trouble again.

Currently, one of his midsized trebuchets sits idly on a flatbed trailer in back of his pole shed, protected from the winter snow by an interweaving of large canvas tarps. If he wanted to, Mullen could lay siege to his neighbor's property. So far, though, he has not declared a suburban war.

And he probably won't—for a while, anyway—as Mullen appears to be a peaceful man. He's a born-and-bred southerner with an easy smile and a soft-spoken demeanor. Although he lives 6 miles down a dirt road in rural Virginia, he's no hillbilly. He's also *Colonel* Mullen, a professor of engineering at the Virginia Military Institute.[3]* As a consequence of his position at VMI, Mullen has been lucky enough to teach both civil engineering, his official job, and timber-framing skills, his unofficial one.

---

*Like the catapult itself, the Virginia Military Institute is sort of a paradox. It is timeless to some, outdated to others. It venerates the power of the military, yet only a minority of its graduates actually go into the military. Like the catapult, many VMI soldiers are built, but few are ever used in an actual war. It is a mix of the most modern educational techniques and the most outdated cultural sensibilities.

VMI is a military college, funded and run by the state of Virginia. Its military history includes a defense of the campus itself during the U.S. Civil War; the VMI cadets fought on the side of the Confederacy. During the Battle of New Market, a group of young VMI cadets fought the Union troops with legendary bravery and valor that is now remembered on a level approaching myth. Fifty-seven VMI cadets were wounded in the battle and ten died. This story of sacrifice has become an iconic part of the school's history, the defining moment in the VMI military heritage.

There aren't many state-run military schools still around. South Carolina has the Citadel, Vermont has Norwich University, and there are military aspects still present, at least to a limited degree, at places such as Texas A&M and Virginia Tech. But these state schools, unlike West Point or Annapolis, have no formal ties to any particular branch of the military.

While they are here, the students at VMI follow a strict military lifestyle, wearing uniforms, living in barracks with up to five to a room, eating at the mess, calling their instructors

At least once a year, he hosts a convocation of the woodiest of the extreme tinkerers—timber framers. These are usually specialized house builders, a type of carpenter who builds structures using wooden joinery instead of metal fasteners. Such building techniques translate to more than just house frames. Mullen and his minions have built a full-sized reproduction of a medieval English catapult.

Building sessions occur annually each spring. A company of VMI students, male and female, sets up a bivouac on the Mullen farm, taking shelter wherever they can find it—in a shed, in the barn, in the back forty in a sleeping bag. They volunteer to spend a long weekend here learning the ancient craft of timber framing.

Experienced timber framers come in from around the world to teach their techniques. Projects have included buildings, sheds, and other structures. But hands down, the most captivating student projects have been a pair of tall, hardwood-beam-construction catapults, ones with massive wooden levers built so that the long,

---

"Major" or "Colonel," and most of all willingly and voluntarily following rules that seem either quaint or ridiculous from the student viewpoint at almost any other college or university. The cadets, as the students here are called, all have closely cropped hair and follow traditional rules of military conduct and honor. They are subject to inspections at all times, and they do things such as constantly assemble in formations for a multitude of reasons. They march as a group to meals at appointed times, they march to chapel, and they march to the barracks to go to sleep.

There are several curricula, but the most popular major at VMI is engineering. VMI graduates a lot of engineers. The faculty are ostensibly military officers, not in the U.S. armed forces, but in the Virginia State Militia. Professors are addressed as "colonel." Their military commission comes from Virginia's governor, not the federal government. Apparently, there is some sort of loose affiliation between the Virginia National Guard and the faculty at VMI, but it's hard to imagine the dire circumstances that would require the academics at VMI to pick up a rifle and leave for active military duty.

The cadets here, engineers and otherwise, seem to accept, even enjoy, the strict rules under which they live and work. Although there is always a fair amount of complaining about the rules, and circumventing these rules is a tradition of sorts, the military "rat" culture (as the system of self-governed groups of student regiments is called) under which VMI life operates is the whole reason for its existence.

strong throwing arm tosses its Great Dane–sized projectiles up and out with grace and a winding, lengthy acceleration.

Several years ago, a visiting history professor asked the VMI engineering department to build a catapult to test some of his theories regarding medieval construction techniques and their relation to catapult building. Mullen and a colleague accepted the challenge, and from there Mullen became an expert in the construction of large medieval hurling machines.

In the past few years, the cadets have helped build the two big catapults that have made VMI's reputation as a catapultin' type of place. Some of the participants, students who have graduated, return often to the Mullen farm to enjoy the spirit of cadet comradeship and pass on what they learned in terms of the technology of such joinery and rigging to the newer students.

Those VMI catapults are finely hewn contraptions, massive yet intricate, and their oak beams are interlaced with complex angled wooden tongues and sockets. The construction is similar to the catapults of twelfth-century Crusader princes, with complicated but strong mortise-and-tenon joints fashioned with hand tools, seasoned oak lumber, and the labor of scores of young military men (and a few women) working under the direction of the master of engines—the "engynour." By learning these timber-framing skills, the cadets will have the ability to improvise their own siege engines, should their Bradley fighting vehicles and Abrams tanks fail someday on a far-off desert.

> > >

It's time for your second shot. This time you let Dean pick out the projectile from a pile of approved pumpkins set out by the event's organizers. All of them are 8 to 10 pounds, hard-shelled, ghost-white pumpkins. Dean picks a pumpkin that's slightly lopsided but should fit the shape of the centrifugal catapult's sling quite well. It seems firm and hard, and with some luck it will remain in one piece until it hits the ground a couple thousand feet downrange in the soybean field.

Another teammate, Janie, takes the pumpkin and places it in the sling. Your centrifugal slinger is similar to the other three; all of them share certain basic characteristics. All of the centrifugals here are three-to-four-story-tall steel towers, each with a large diesel engine, so big it may have been scavenged from a deuce-and-a-half truck. The engine is either mounted at the top of the tower or left at the bottom and joined to the rotor via a couple of U-joints and a drive shaft. Either way, attached to the engine's drive shaft is a very long, carefully counterbalanced crossarm, with a pumpkin-holding sling at one end.

To start, a team member inserts a pumpkin in the sling and then backs away while another teammate starts the diesel engine. Slowly at first, the diesel engine chugs and smokes, and the crossarm starts to spin, gradually building up speed, faster and faster, until the arm dissolves into a propeller-like blur. As it spins, a breathy, whirring sound grows louder until the noise can be heard pulsating throughout the length of the half-mile-long firing line, and maybe as far away as Millsboro.

Indeed, the rotor noise is very loud. Everywhere, all eyes automatically turn toward the wildly spinning Centrifugal. It takes a while to come up to top speed. Finally, with the crossarm spinning at the G-force limit of what you think pumpkin and machine can withstand, you slam down hard on a lever at the instant when you feel the release angle is closest to perfect. This releases the sling, and the pumpkin hurls skyward, tracing out an arc of the longest parabola that anybody here has ever seen. Your timing was good, you think—not perfect, perhaps, but darn good. The pumpkin remains airborne for what seems like a half a minute, going so far that it becomes invisible to unaided eyes, and it is only findable again when the pumpkin splats into shards of organic confetti way, way out in the field.

The centrifugal catapult, when deprived of its 10-pound pumpkin counterweight, becomes seriously unbalanced after firing, and the machine dances clumsily on its base until the arm slows down and stops.

The spectators crowded around the long snow fence that was erected to separate them from the pumpkin chunkers and the dangers of their moving machine parts are ecstatic. Clapping and toasting the throw, they raise their beer cups, after which the congregants mosey farther along the fence line to watch the next chunking activity.

This was a very good toss, and there is much rejoicing by all. Dean and Janie and the others are very pleased. But the joy is short-lived, for on the very next throw, just a few minutes later, one of your competitors sends a pumpkin about 100 feet farther. It's not yet time to panic, though; there is still one more toss to go.

ON DOWN THE FIRING LINE

Besides the centrifugals, there are a lot of other pumpkin-hurling technologies present at the World Championship Punkin Chunk. Walk down the firing line and you'll find air cannons, some of the smaller ones built by youths seventeen and under whose craftsmanship rivals, and in a few cases surpasses, the work of their parents back up the firing line. Pumpkin chunking is a family affair for quite a few, and it is not uncommon for three generations in one family to work together year after year on a machine. After a while, the younger ones go off on their own, and they start here, in the youth division.

Next on down the line are the spring-loaded catapults. The twenty or so catapults lined up next to one another all work similarly, but many have their own twenty-first-century take on second-century technology. Instead of a springy bundle of twisted cords made from the neck tendons of a bull, as the ancient Romans made theirs, the pumpkin-chunking catapults use garage-door springs—lots and lots of garage-door springs. Here, the typical catapult builder welds together a big steel contraption with room to attach up to forty extra-long door springs. The plate to which the springs are attached is then connected to a pumpkin-holding arm in such a manner that when the device is retracted using a very large hydraulic cylinder, the springs extend and the pumpkin is held in firing position. Upon release, which is accomplished by yanking on a rope attached to the trigger, the steel springs exert a mighty tug on the swing arm, and this sends the pumpkin flying, often to distances approaching a half mile.

The gravity-operated catapults, trebuchets, are usually more modest affairs, most consisting of a 10-to-20-foot-tall steel superstructure, a tall lever arm, and a counterweight made from concrete or scrap iron. Evidently, despite the ancientness of the technology, these are enormously tricky devices to operate successfully. One of the trebuchets is made from a colossal boom-attached swinging arm, powered by a combination of counterweights and a parallel array of very stiff torsion springs. The device's builders insist that the machine was built in accordance with some ancient numerological, spiritual, and feng-shui-like guidelines, but they do not quite have its yin and yang in perfect harmony. Sometimes the trebuchet behaves perfectly and heaves the projectile hundreds of feet downrange into the soybean field, inducing the machine's builders to jump around gleefully and high-five one another. Usually, though, the thing tosses the pumpkin 200 feet straight up and the pumpkin lands either right on the trebuchet or maybe a few feet in front of it. And once in a while, say every fifth or sixth toss, the thing winds up, the boom swings down, and it somehow tosses the pumpkin more than 200 feet in the exact opposite direction of where it was aimed,

THE TECHNOLOGY OF CENTRIFUGAL CATAPULTS

In a typical Roman-style catapult, a long throwing arm is inserted into a cocked torsion spring. When the arm is released, it spins around its axis and any projectile, whether it's a rock, spear, or pumpkin, is hurled out when the gravity holding it to the throwing arm is overcome. The throwing arm of a typical catapult remains in contact with the projectile for only a fraction of a single revolution of the arm. At some point, the gravitational forces that keep the restraining pouch bound to the whirling arm are overcome and the projectile soars off, ideally toward its target.

A centrifugal catapult is designed differently. The restraint holding the projectile to the throwing arm is mechanically controlled by an operator, and so its release is not preordained by gravity and geometry. When the operator decides the time is right—which is always after a long series of constantly accelerating revolutions of the spinning throwing arm—then and only then is the projectile released. Tremendous amounts of angular momentum are built up as the centrifugal catapult's angular velocity grows exponentially due to the work of the attached piston-driven engine.*

A typical centrifugal catapult consists of just a few basic parts:

---

*Ordinary momentum is a measure of a thing's natural predisposition to move at constant speed along a straight line. The more mass and speed a thing has, the greater is its tendency to continue on the same straight line.

An automobile moving at 55 mph has more momentum than a bicyclist moving at the same speed, because it has much more mass. The car hitting a wall at 5 mph does not cause as much damage as it does at 50 mph because it has one-tenth the momentum. For things moving in straight lines,

Momentum = Mass × Velocity

Some items, such as pumpkins thrown from the slings of centrifugal catapults, move in a curved path. The momentum of curvilinear movement is best described by

the throwing arm, the release, the engine, and the tower. A simple device in concept, but what a challenge to build successfully.

It is a straightforward enough design task to build a steel and iron superstructure. Given a good ironworker's ability to weld steel and bolt together angle irons and I-beams, this poses no particular challenge. Building a counterweighted throwing arm and connecting it to a surplus gasoline or diesel engine is a considerably more difficult job, though not overwhelmingly so. But building a good release mechanism can be complicated. The centrifugal catapult needs a mechanism that reliably launches the pumpkin at just the right time, at just the right angle, and at just the right speed. This is, engineering-wise, a considerably bigger job. Release too early and the pumpkin goes straight up. Release too late and risk pumpkin shrapnel from an ugly splat a few feet downrange.

---

a property called "angular momentum," which accounts for the curved motion of the projectile.

Angular Momentum = Mass × Velocity × Distance (from the point that the object is orbiting)

Engineers are quick to exploit the laws of physics to their advantage, and here the laws of physics provide a lot to exploit. Newtonian physics tells the clever pumpkin chunker that the amount of energy imparted to the pumpkin projectile is related to the square of the rotational velocity of the arm. The mathematical formula that describes the relationship between the speed of the rotating arm and the energy imparted to the pumpkin is a simple one, but fraught with importance to the pumpkin-throwing engineer:

Energy of a Spinning Projectile = $\frac{1}{2}I \times \omega^2$

where I is a non-varying quantity called the moment of inertia and ω is the rotational velocity of the pumpkin.

Understanding what a moment of inertia is or how rotational velocity differs from regular velocity is not important for the casual observer. What is important is to know that the energy imparted to a soon-to-be-launched pumpkin exponentially increases with the speed of the throwing arm's rotation.

For example, an arm spinning at two revolutions per minute will impart four times the energy of an arm spinning at one revolution per minute. And an arm spinning at four revolutions per minute will have sixteen times the energy. So as the long arm spins faster and faster, the energy available to propel a pumpkin grows enormously.

When the engine is engaged, the arm will travel smoothly in graceful, whooshing circles, slowly spinning up like the mammoth propellers of 1960s-era Soviet cargo planes—that is, as long as the weights on both sides of the throwing arm are very close to equal. The chugging, smoking, huffing engine spins the arm faster and faster, increasing the energy of the system with each rotation. At just the right time, the release mechanism is engaged and the pumpkin flies off, a speck against the sky, soaring toward the edges of the field and its glorious destiny of ruin.

And what happens to the centrifugal catapult after the pumpkin projectile release? As soon as the release mechanism is engaged and the pumpkin discharged in flight, the catapult itself experiences a rapid fall from grace. With the pumpkin gone, the whole contraption is overwhelmingly and completely thrown out of rotational equilibrium, the whirling arm being 10 pounds short of any condition approaching or resembling balance and order.

A measly 10 pounds doesn't seem like much on a machine this size, except when a 30-foot-long rotating member is spinning at speeds approaching three times those of an LP record. Nature abhors such awkwardness. Great forces instantly present themselves on the bearings holding the shaft, the structural members of the support tower, and the footings anchoring the catapult to the ground. The catapult strains mightily to regain its inner harmony, attempting to shift its rotational center back to a point more in line with its new and radically altered center of inertia. The framework staggers on its foundation like a drunken giraffe. The unbalanced forces stress and strain every part of the device, causing it to twist and rock, groan and squeak, until the arm gradually slows and then stops rotating. Eventually, the giraffe sobers up and stands tall and quietly again, and it's the next machine's turn in the spotlight. <

which generally puts the pumpkin in the vicinity of some poor unfortunate's Chevy in the parking lot.

Grigg Mullen's VMI colleague Wayne Neel is perhaps the dean of catapult scholarship, at least on the western side of the Atlantic. For Colonel Neel really understands the physics and mechanics of catapults. He has mathematically modeled them and understands them on a deep scientific and theoretical basis. But he also understands them on a personal, practical, and nonacademic level.

All over the world, there are Technology Underground enthusiasts who design catapults authentically, using the best building methods and materials peculiar to the thirteenth century. To the extreme tinkerer who loves siege engines, such machines represent the pinnacle of manual craftsmanship, historical scholarship, and engineering sans modern math and physics. Catapults are machines that can be built without the need for any level of math past arithmetic, and practically without the need for any knowledge of applied physics. Any machine that can be constructed wholly from basic hand tools, yet is 60 feet high and offers its creators the power to knock down stone walls 12 feet thick, is certainly interesting.

Neel and Mullen's largest siege engines are trebuchets. At first blush, they appear to be very simple in concept—a long, strong wooden timber is affixed to a framework through a single fulcrum. At one end of the lever arm is a sling that holds the projectile (say, a large rock, a barrel of flaming oil, or a diseased cow carcass), and the other end holds a counterweight, usually a big and heavy box of rocks. The counterweight is lifted by a group of men and animals operating winches. Upon the order of the engine commander, the counterweight is released and the swinging arm flings the projectile in a high, looping arc.

Throwing a 100-pound ball of rock a distance of 200 yards

requires counterweights of many tons and structural timbers several feet thick. The joints are made with mortises and tenons, and the entire machine is made solid and sturdy through the use of locking hardwood pegs and wedges.

Mullen and Neel and their cadre of timber framers have designed and built a total of five large, authentic, and painstakingly crafted siege weapons to date. The first two were wooden-framed trebuchets and worked quite well. They were built to add to the body of academic knowledge relating to twelfth-century war machine construction, using the same types of techniques and materials that an English carpenter might have used during that time period. The machines were built with VMI cadet labor and master timber framer supervision. The wood for the trebuchets was hewn from great beams of oak and pine. Each beam and each supporting piece was connected to the whole not by nails or modern fasteners but by finely crafted joints—mortise and tenon, splined, dovetail. The joints were cut by hand in the wood using traditional methods that required cutting, sawing, and chiseling, using mostly hand tools and hour upon hour of hand labor. These trebuchets were built the way a siege engine of the 1300s would have been.

The trebuchets worked well, tossing all sorts of items, heavy and light, round and square, as desired by their makers. They hurled far and true and they continue to do so, earning a certain amount of notoriety at VMI events—football games and such—by flinging stuff into the crowd. Ultimately, one of the big hurling engines found profitable employment touring the country for a large brewery, flinging barrels of beer as a publicity stunt.

Mullen and Neel's third project was a massive double-spring-powered rock thrower, made for an English television show with the help of native English timber framers. This machine was called a *ballista,* and the giant slingshot stood three stories high when completed. The idea was to authentically re-create the feared stone thrower of the first-century Roman legions. The project was funded by PBS, and the resulting television show was a popular *NOVA* series episode called "Secrets of Lost Empires."

Then the two engineers got together again for a fourth siege project, which was a BBC-sponsored effort. This was to be an accurate mechanical rendering and construction of the famous (but never built) stone-throwing crossbow of Leonardo da Vinci. According to the dimensions found in a Leonardo manuscript called the *Codex Atlanticus,* this design called for a gigantic bow, 25 yards across from bow tip to bow tip.

The medieval Italians never built the stone-throwing crossbow, for a good reason: Despite the best efforts of the modern team's legion of expert timber framers, engineers, academics, and television producers, the giant crossbow could do no more than feebly fling stones with less power and range than that of a so-so baseball pitcher.

Colonel Neel is a tenured engineering professor at Virginia Military Institute who in a certain light and from a certain aspect bears a resemblance to Christopher Walken. His corner office, which overlooks the football field, is in the lower reaches of the engineering building, a gray-green fortress-like structure similar to all the other buildings on the campus.

The office is full of books on catapults: copies of illustrated manuscripts from the libraries of the Middle Ages, books on Chinese science and technology, and many books of ancient drawings and carvings.

"There's still a lot to be discovered, to learn about them," he says. "And I think short of some unexpected, and to be honest, lucky archeological find, the best way to learn more is to do accurate mechanical reconstructions and apply the laws of engineering to them—to figure out what makes sense and what works from a building standpoint. That's why building trebuchets and *ballistae* is a meaningful exercise from both scientific and historical academic perspectives."

No one has ever seen a real ancient catapult, one built before the mid-1400s, after which gunpowder cannons were used exclusively. There are no excavated examples of a catapult, save for a few decayed *ballista* parts dug up in Spain and Israel. But Neel believes he knows what a trebuchet or *ballista* really looked like. According to him, they didn't look too much like the ones typically shown in books and movies.

"The way catapults are typically depicted in the movies or in pictures . . . well, there are many inaccuracies there. There is some germ of truth, but basically, real catapults were much different. They had to be, because the Hollywood kind won't work, not under real siege-like conditions." Neel picks out a model catapult from his desktop and continues, "The way most people think of a catapult is a large wooden frame with a tightly twisted coil of rope running from one side to the other. There is a big arm with a cup at one end, which flings rocks and such when the arm is retracted and then released. The arm spins violently, hurls the rocks, and then is abruptly brought to a halt by hitting a padded crossarm attached to the frame.

"There are lots of inconsistencies and inaccuracies with that depiction," he says. "The ancient writers, the Greek and Roman engineers, don't say anything about twisting the cords. They were pulled tight but not twisted. That's a significant difference. And you know how people think that catapults always have that horizontal crossarm that the throwing arm bangs into? I don't think they really did, because if the throwing arm impacts a rigid crossarm every time it fires, the arm is simply going to break.

"In reality, catapults were usually long, flat affairs with multiple tightened skeins of ropes, not coils, and they were pulled tight and then tuned by sound, much like a modern piano tuner, to make each cord's tensioning consistent. It was a complicated and intricate sort of craftsmanship. Catapults were not something just any soldier or house builder could make."

Neel lowers his voice and speaks as if he is going to reveal an important secret. And indeed, that is what he does.

"The secret," he says, "to the power of the catapult is the way the ancients affixed the projectile to the throwing arm." He explains in detail how they used a flexible rope with a pouch. One end they fixed to the middle of the throwing arm and the other they tied in a sliding loop.

Neel explains further that a real catapult always used a rope sling, never a cup, to hold a projectile. The whole idea behind a catapult is to use leverage and weight to throw something big a long way. A very long throwing arm is the way to make the machine throw projectiles farther and faster. A trebuchet can only be built so tall and so high, however, and that limits how long the throwing arm can be. But by adding a flexible sling to the end of the throwing arm, it whips the projectile around as if the arm were twice as long. The sling makes it twice—maybe three times—as powerful. The sling is the secret behind the power of the machine.

CONFEDERATE AND YANKEE CHUNKERS

Waddy Thompson is a tinkerer's tinkerer. His family has resided in south Florida for five or six generations. In fact, his grandfather worked on the first telegraph line in south Florida and was the man who first relayed the message that the U.S. warship *Maine* was blown up in Havana's harbor. In Waddy's lifetime, the Thompsons have seen the rural farmland around their home near Fort Myers disappear under wave after wave of housing developments and commercial sprawl. Before condos and housing developments took over, there was a lot of agriculture in this area, and thus an odd connection to pumpkin chunking.

The innate desire to see things go boom, whoosh, or splat has a long history among Thompson's friends and family. Prior to building large vegetable-shooting air cannons, watermelon roulette was a favorite pastime at barbecues and parties. In south Florida, farmers can grow three crops of melons a year. They are ubiquitous and cheap, and this led to the evolution of the uniquely south Florida pastime of gambling with large fruit.

To participate in this activity, one buys a china dinner plate from the event organizer for, say, five dollars. Then the plate owner writes his name on it, marches out into a large field, and sets the plate down in a spot that he considers lucky. After all of the plates are positioned, a biplane takes off from a nearby airfield with a cargo of large, ripe, Florida-grown watermelons.

As the plane flies over the field, the co-pilot randomly tosses a watermelon out the door. If the watermelon misses smashing a plate, then the pilot circles the field and releases another melon. At some point, due to the laws of probability and statistics, a melon makes a direct hit on a dinner plate and that plate's owner goes home a few hundred dollars richer, taking back his entry stake and everybody else's as well.

While a biplane on a low-altitude bombing run of a Florida field dropping unguided melons is interesting, it is an expensive and possibly illegal way to accomplish the task of gambling with produce. Perhaps, Thompson thought, artillery of some sort might be a better way to accomplish this. He set to work inventing an air gun, one with the wherewithal and breech diameter to send a 15-pound Royal Sweet watermelon in a high parabolic arc toward a field planted with cheap dinner plates. Obviously a machine that can shoot melons would be similar to one that shoots pumpkins. The melon shooter led to a pumpkin shooter, and here he is. In fact, Waddy notes, a pumpkin is probably easier to shoot since its rind is harder.

Waddy Thompson works on the edge of Guy Culture technology. He is an advocate of the kind of slow-and-steady, trial-and-error building techniques that win races. As Thompson says, "Building this kind of stuff isn't a matter of sitting down and drawing a complete machine on a computer or piece of paper, then building it to the design. It actually took us about two and a half years to work out all the details for the gun.

"The materials that it was built from are almost all donated from scrap piles or purchased from scrap yards. The only things that were bought new for it are the main valve and valves for the air pump, and a coupling used to assemble the two sections of the fiberglass sleeve. I figure that we have less than $500 cash in the whole thing. The biggest expense is getting it and us to places like the World Championship Punkin Chunk in Delaware. This year my family spent about fifteen hundred dollars on the trip for five of us.

"The time spent is not a factor in this sport. When you see people with their jaws dropped around their knees when they witness a good melon shot, it's as good as any hole-in-one for a golfer to me. When people ask why we launch pumpkins and melons, I ask them, why do people play golf? At least we don't have to chase a little ball around after we launch. The most common remark that we hear about the sport is: 'You guys have too much time on your hands.' Well, the same goes for you golfers."

This sentiment was widely echoed by many of the builders. New Hampshire's Randy Kezar, a member of the engineering school of chunker building, doesn't care a whit what others think of how he spends his time. He doesn't even care too much about how far his machine tosses the pumpkins. He's got his own reasons for being here.

Kezar's pumpkin shooter is called the Honeydew Screw and is basically a 24-foot length of aluminum pipe welded to a 300-gallon propane tank. The whole thing is bolted, welded, and glued to the back of an old, beaten-up gray station wagon. Kezar works as an engineer for an auto parts manufacturer up near Nashua, and he's an uncommonly interesting fellow to engage in a technical conversation.

How important is getting good distance to Kezar? He says, "I forget who the architect was who stated something like 'form follows function.'* That is certainly part of it, though I don't get too wound up with the distance. At this contest I am there mostly for the party. My other objective is to give the town zoning board back home a 'tune-up.' Pumpkin-hurling machines do not fit well in the municipal zoning code. The zoning folks get bent out of shape about used cars, a house that needs painting, someone who hasn't mowed his lawn, and so forth. My pumpkin shooter is good for a grin or two when my neighbors drive by."

So why do it?

"The idea of tinkering and pumpkin chunking—all of the effort we place into something as, well, sophomoric as this," he says, pausing to choose his words carefully, "is hard to justify in terms of dollars."

But he's willing to give it a try.

"In its fundamental sense, pumpkin chunking is an idiosyncratic hobby and a wonderful excuse to party. Hobbies, I think, are something I see becoming increasingly less common in mainstream culture, what with the advent of intoxicatingly realistic video games, television, and the Internet.

"Justifying this is like asking an avid gardener to explain the rush he gets from a pretty laid-back endeavor. There are those who find serenity in things as seemingly mundane as mowing the lawn every two days in the summer, like my neighbor does.

"Working in the garden is very similar to working on my pumpkin chunker. I love the long evenings of summer leaning on the cannon, fixing it up. My girlfriend describes this as a bunch of guys standing around trying to decide which machine bit belongs where."

To competitors, the bottom line is that pumpkin chunking doesn't require rationalization. It's worthwhile simply because they like it.

---

*The quote is generally attributed to Chicago-based architect Louis Sullivan.

> > >

Waddy Thompson's pumpkin shooter, which he named the Spooky Bazooky, is a marvel of extreme tinkering. While lots of people have built pumpkin-hurling devices with as much or more power, none of them equals the Spooky Bazooky in terms of ergonomics and design sophistication. The Spooky Bazooky may be to pumpkin tossing what the Stealth bomber is to aerial warfare: It isn't the biggest, but it has the most elegant engineering and does what it does unlike any other.

Consider the way the Bazooky works. While other big-vegetable shooters use large diesel engines or gasoline-fired compressors to provide the motive power for their shooters, the Bazooky uses an ingeniously modified bicycle mounted on top of the gun to turn a shaft. This shaft in turn operates three double-acting pneumatic cylinders linked together like the action on an upright piano. As the operator pedals away, the cylinders move up and down, filling the storage tank with compressed gas. As the pressure inside the storage tank builds, so does the back-pressure, and the bicycle gets harder to pedal.

To overcome this, Thompson designed a motorcycle transmission to allow the rider to downshift and thereby continue to increase tank pressure despite the mounting air resistance. Using nothing but the energy derived from human muscle power, this gun shoots melons and pumpkins well over half a mile, farther than many of the "championship"-caliber contestants can shoot with their chugging, fossil-fuel-powered air compressors and truck engines.

Further, the human-powered chunker that Thompson and his son designed has an ingenious lever-operated, multi-bar mechanical loader that allows a single man to pull a lever to open the breech, drop in a pumpkin, and close the breech, all within about thirty seconds. The Bazooky has an inlet to allow the device to operate off compressed gas as well as the bicycle pump. If hooked to a large compressed-air source, their gun fires at about the same speed as a major-league baseball pitcher throws. Compared to the lengthy loading-and-firing intervals that the other guns require, it's something of a technical marvel.

> > >

It's time for your third and final try. This time the pumpkin choice was a team decision, and it was a tough one, since the remaining pumpkins were a little picked over. Still, the 9-pounder you all agreed upon does seem acceptably hard and aerodynamic.

Dean places the pumpkin in the sling and Janie starts up the diesel for the last time. Black smoke chugs from the engine exhaust. The rotors start to make their beating noise: *whoomp . . . whoomp . . . whoomp . . . whoomp whoom woowowo.* Then the rotor noise becomes a continuous whir.

"How fast is it going?" shouts Dean over the noise.

The noise is tremendous. "I'm not sure. Maybe a hundred rpm . . . maybe more."

The rotor has never spun so fast. It's all holding together, but just barely. It's time to let the pumpkin fly. With both hands you grab the release lever tightly. If you time it just right, that pumpkin will go a mile. You jerk the release arm.

You look downrange and see . . . nothing. Nothing but a few chunks of rind and pumpkin skin, seeds, and white pumpkin flesh dripping from the sling. It's pie time again.

6. air GUNS

New York City's Central Park is crowded with people on this warm spring afternoon. Some are lying around; others are sitting, running, walking, or tossing balls or Frisbees. You are standing in Sheep Meadow, a large expanse of semiflat open land near the southern end of the park. Although it is still late March, the day is warm enough and pleasant enough that many people have taken their jackets off and rolled up their shirtsleeves.

You are not here to loll about or walk the footpaths. The plan is to find a spot that is open yet secluded and not chock-full of people. That's a tough order to fill because this is Central Park, and on a day like this, there are people just about everywhere. But still, you and your entourage seek a quiet corner, for in the long fabric bag that you've been lugging all the way from the Columbus Circle subway station is an item of questionable legality. Well, actually, that's not true. You are sure that it is, in fact, not legal. In the bag is your flint-and-steel-ignited, antiperspirant-powered, potato-shooting air gun.

The leader of your team is Harlem resident Todd Keithley, a Yale-educated media consultant. Keithley, a well-spoken, friendly fellow in his mid- to late twenties, with a slight build and close-cropped hair, walks this way and then that way, reconnoitering the downtown end of the park, until he finally finds a promising spot in the southwestern part of the meadow, near Tavern on the Green. Yes, you and Todd and the rest of the group decide, this probably is as good a spot as you'll find. It's not exactly private, but it is kind of secluded, and at least at the moment there are no cops in sight.

You unzip the bag and take out your potato cannon, aka the Spudzooka, the Starch Accelerator, or the Gun that Abuses Vegetables. It is something you, Todd, and another odd but knowledgeable acquaintance made yourselves the day before, a contraption of polyvinyl chloride pipe, plastic connectors, and a few assorted pieces of camping equipment. It doesn't appear to be dangerous, but after only a few minutes of setup, this device made from modified plumbing supplies will be able to blast a potato across several avenues with powerful acceleration and an admittedly dangerous muzzle velocity.

After assembly, Keithley takes a nice big Idaho and jams it down into the barrel of the gun, the sharp beveled edge of the PVC pipe cutting the potato into a smooth plug of starch. Down it goes, pushed about 24 inches into the barrel with a ramrod made from a wooden dowel. Another person holds the gun on her shoulder while you unscrew the cap from the back end, where the spark igniter from a Coleman stove has been cleverly modified and inserted.

The spark igniter is a mechanism composed of flint and steel. When a knurled knob is turned, a spark is created inside the gun barrel, which will ignite whatever flammable compound happens to be inside. When the flammable compound explodes (actually it doesn't explode per se but "conflagrates," although this is a technical point), the hard PVC sides of the gun do not yield to the pressure. But the potato does yield. In fact, it yields big-time. It will be pushed out of the long gun barrel at amazing velocity, courtesy of a blast of rapidly expanding gas.

The time has come for the shot, and this will be the dicey part and may attract unwanted attention, for the spud gun is quite a noisy item. As the gun holder steels herself for the expected recoil, you unscrew the cap to the combustion chamber and charge the gun with a two-and-a-half-second burst of Right Guard antiperspirant, which is mostly an aerosol mixture of butane and propane. You seal the end and move your hand to flick the trigger.

But as you start to turn it, your brain commands your hand to freeze. For coming off the footpath to the right, you suddenly notice a flash of reflected sunlight, glinting off the helmets of two of New York's Finest, in blue uniforms, mounted on horseback.

> > >

Air guns are simply gun barrels that use air pressure from a pressure reservoir instead of a powder charge to shoot a projectile. Air guns are much older than gunpowder weapons and have been used for millennia. There is archeological proof that lung-powered blowguns were in use around A.D. 100. But even without concrete evidence, there is little doubt that air guns in some form were around for hundreds of years before that.

The earliest mechanically powered air guns were simply blowguns that used a bellows attached to the breech. Instead of huffing into the pipe, some clever northern European came up with the notion of putting a squeezable bag on the end. When the bag was squeezed, the compressed air shot a dart or pellet out of the gun. And if the bag was squeezed using the mechanical advantage derived from a system of levers (imagine a fireplace bellows), then the gun could be made to shoot much more powerfully than could be accomplished through simple lung power alone.

The oldest existing mechanical air gun is thought to be a specimen in the collection of the Livrustkammar Museum in Stockholm, which the museum dates to about 1580. Inside this old gun, a spring mechanism operated an air bellows located in the stock of the gun when the shooter pulled on the gun's trigger. The spring caused the bellows to forcefully exude a powerful air gush, which shot a specially shaped bolt, or dart, toward the intended target.[1]

By the early seventeenth century, air-powered darts were being shot across Europe for sport in a variety of ways. According to the people who study such arcana, spring-powered air guns activated by a moving piston (which was a big improvement on the earlier bellows-reservoir technology) quickly appeared. The first detailed description of such an air gun is found in the *Élémens de l'Artillerie* by David Rivault, who was preceptor to Louis XIII of France. He ascribes the invention to a man identified only as "Marin, a burgher of Lisieux," who presented the first air gun to England's Henry IV. By the turn of the nineteenth century, air guns had developed to the point where they were likely more accurate and more powerful than contemporary black-powder weapons of similar size.[2]

Circa 1800, air guns of this caliber and quality were expensive to make. It took a great deal of time, knowledge, and precision to make a device of this type, because the components—valves, locks, cylinders, and reservoirs—had to be very carefully machined. Consequently, they cost far more than a simple black-powder rifle and therefore were beyond the reach of most people. Additionally, there were lots of things that could go wrong—for instance, pressure leaks, exploding reservoirs, and sticking valves.

Still, for those who could afford them, air guns offered many advantages. For one thing, they could be fired rapidly—several times a minute, which was far more quickly than the muskets of those days, which required a load, tamp, and fire procedure. By comparison with a smooth-bore muzzle-loading musket, air guns were veritable machine guns.

Second, they didn't expel a lot of smoke. This made it easier to aim the next shot if it was needed, for the line of vision wasn't obscured. Third, a shooter didn't need to be concerned about keeping his powder dry—the gun worked as well in damp as in dry conditions.

The most historically important American air gun was a rifle carried by Meriwether Lewis during the Lewis and Clark expedition of 1803–06. The actual gun (although there is some controversy surrounding its pedigree) resides in the Virginia Military Institute's museum of historical weapons. The VMI museum claims that the .31 caliber, flintlock-style pneumatic rifle in its collection is the one built by expert clockmaker Isaiah Lukens in Philadelphia. Lukens gave the gun to Lewis prior to the Corps of Discovery expedition to the Pacific Northwest. The butt of the Lewis rifle is actually a metal canister designed to contain a charge of very-high-pressure air—between 700 and 900 pounds per square inch.*

When the trigger of the pneumatic rifle is pulled, a charge of air is metered through channels and valves, coursing its way from the reservoir in the gunstock end to the bullet chamber. When fired, the round leaves the barrel with a loud whooshing sound and a very high velocity. Such a gun could easily take down a full-grown deer. According to the records kept by the Corps of Discovery, the air gun did its job well. It was effective for both hunting and for impressing the Indians with its power and novelty.

Perhaps the high point in air gun development culminated with the James Bond–like "air cane" craze of late-nineteenth-century England. These walking sticks were de rigueur for the upper-crust English gentleman of the day. Although an air cane appeared to be but a simple walking stick, inside it was everything needed to shoot a large-caliber bullet with a very peppy charge of compressed air. Truly, the air cane was a dangerous weapon and an impressive means of self-defense for the security-minded Victorian Englishman. They fired with a power equivalent to that of a modern revolver.

*That's a lot of pressure. A car tire carries a pressure of 35 pounds per square inch.

THE GUNS OF DELAWARE

Like the builders of centrifugal catapults, backyard technology enthusiasts whose interest tends toward gargantuan pneumatics think of the World Championship Punkin Chunk as the Super Bowl and Daytona 500 rolled into one. It is an event, though, that by most lights could be considered fairly ridiculous. It is difficult to find great social and scientific significance in this activity. But is it just a grand joke, like TV's *Seinfeld,* a show about nothing?

According to the participants, organizers, and fans, no. The participants, at least some of them, are extremely serious in their bid to win the Punkin Chunk trophy. A few teams spend upward of $40,000 on their air guns, and that kind of money breeds some intense feelings. There is tough competition between the teams, of course, and there is also rivalry between groups.

There are two general divisions, or schools of thought, regarding builders and their credentials. The first are the good-old-boy teams.* These are the people who originated the event and have been there from its beginnings in the 1980s. These teams are made up of local fellows, most of them rural Delawareans from places such as Lewes, Milton, and Seaford, and their years of on-the-job experience welding, rigging, and piping are utilized to the utmost advantage on the big air cannons. The good old boys don't make mathematical calculations and don't use formulae from *Machinery's Handbook* to build their guns. Instead, they depend on the twin mechanical design philosophies of T-LAR (That Looks About Right) and overkill to build big-time air gun pumpkin chunkers. It's cut, cut, weld, weld, pound, pound, and after a while a pretty good machine gets built from recycled ammonia refrigeration tanks, purloined irrigation pipe, and salvaged chemical-plant valves. Everything, from the barrel length to the

*The term *boy* fits because the participants are overwhelmingly male. However, there are exceptions. One noteworthy team consists entirely of women.

size of the air tanks to the platform on which the gun sits, appears to be overengineered and oversized. The safety factors built into these guns are large enough to forgive a multitude of design sins.

A lot of these good-old-boy teams are oversized as well in terms of the number of builders, designers, and hangers-on who participate. Take, for example, the team that builds, maintains, and shoots one of the largest guns, the 100-foot-long air cannon called Old Glory. Old Glory first came on the scene in 1997 and consists of a pair of way-too-big propane tanks manifolded together and connected to 70-odd feet of steel gun barrel. A man named Joe "Wolfman" Thomas is ostensibly the leader of this team, but the crew is so big and so loose that it appears no one is really sure how many people are working on the guns. They just know that enough good people will show up in the field with tools so that the gun will get erected on time.

Jim is a typical team member. A boat mechanic by trade, he's willing to use up many hours of his free time laboring over the big gun and its accompanying logistical train, including a 40-foot flatbed trailer and diesel truck. Jim has longish black hair, dirty hands, and an engaging smile. He dresses in workman's coveralls and, unlike most, he never drinks while working on the gun. "Plenty of time for that later," he says. "Now's the time to stay focused on the gun."

Jim and the other Old Glory guys seem to have a personal, anthropomorphized relationship with the gun itself. They bitch about it constantly—"the goddamn barrel's too short," "the goddamn firing angle is all wrong," "the goddamn air tanks are too large"—but deep down they really like it a lot. Even when complaining about it, they sort of lower their voices and turn away so that the gun doesn't hear them and start to feel bad.

The team with builders Pete Hill and Ralph Eschborn represents the flip side of this equation. Tall, thin, and bespectacled, Hill and Eschborn are as different from the good old boys as NASA is from NASCAR. These two are chemical engineers, registered within their states to perform such work, and they meet all of the stringent criteria required to get an engineering license. Like true engineers, they spent as much time on their computers designing their cannon, Big 10-Inch, as they did actually building it. The air gun was computer-modeled, its physical properties analyzed, and its performance enhanced by applying the principles of fluid dynamics, engineering mechanics, welding metallurgy, and shoe repair.

The two chemical engineers seem to understand with amazing clarity what happens deep inside the gun barrel during the first few milliseconds after the pressure is applied. They've made a science out of determining the effect of pressurized air upon pumpkin flesh and can explain in detail the physics and engineering of airborne pumpkin destruction.

The differences between the two teams are as wide as the Chesapeake Bay, but they do also have a lot in common. Both teams have dual goals. First, they want to win. Second, they want to drink a lot of beer. The goal over which they have complete control is usually accomplished early and often.

While good-natured competition is the general order of things, the activity is not without controversy. Once a relative of one of the principals on the Old Glory team accused the Big 10-Inch team of cheating. Stories differ according to whom one talks to, but in the end it was said that they had used compressed helium gas instead of compressed air to propel the pumpkin. Because helium is lighter than air, the escaping helium gas apparently departs the barrel at a higher velocity for a given pressure, and this in turn improves shooting performance. To an engineer, experimenting with incremental improvements and tweaking design parameters is as natural as screwing in a higher-wattage bulb to get more light. But to a T-LAR type of guy, it's nothing but cheating.

Clearly, the rules do state that nothing but compressed air is allowed. The engineers said that they did not know it was against the rules to use helium, and furthermore, they argued, only the single toss in which helium was used should be disqualified. But in the end, the rules are the rules, so the title was awarded to Old Glory on a technicality.

At the World Championship Punkin Chunk, the distance records for pumpkin throwing are under serious pressure. The holy grail of pumpkin throwing is the mile barrier. A mile is a really long way to throw anything. A 10-pound pumpkin is a fairly big object, and throwing something that size ten to twenty city blocks is a task that requires simultaneous application of power and delicacy. While throwing a cannonball such a distance might be small potatoes for a military cannon powered by 30 pounds of explosive cordite, for the homemade air guns on display here it is a mighty big job. Still, the people who compete here are clever and hardworking, and lots of them seem to think they have a shot at the mile benchmark. According to the *Guinness Book of World Records,* the greatest distance that a pumpkin has ever been hurled, whirled, or shot is 4,491 feet. That was accomplished in Morton, Illinois, in 1998 by an air cannon called the Aludium Q-36 Pumpkin Modulator.*

These air guns stand out in the minds of the spectators, even outshining the centrifugal catapults and giant slingshots. There is no doubt that these are the stars of the show. The bigger they are, the more people clamor to see them work, for there's something undeniably interesting about a big gun, and a 100-foot-long gun made for the most part in a garage or driveway is a true testament to an intrepid individual's ability to build industrial-sized projects on a shoestring budget.

The fact that each pumpkin-chunking air gun has a unique

---

*Fans of Warner Brothers cartoons may recognize the name Aludium Q-36 Modulator as inspired by the well-known cartoon "Duck Dodgers in the 24½th Century." Director Chuck Jones created a cartoon spoofing the Buck Rogers movies of the 1930s. This cartoon featured Daffy Duck, Porky Pig, and Marvin the Martian, who owns the Illudium Q-36 Explosive Space Modulator. Many animation lovers consider "Duck Dodgers" to be the greatest and funniest Warner Brothers cartoon of all time.

name—for example, Bad Hair Day, Sky Buster, Please Release Me—and personality brings to mind the famous big guns of historical times, such as the Columbiad (the fictional cannon of Jules Verne that could shoot a spacecraft to the moon), the Basilica (the huge Turkish cannon that breached the walls of Constantinople in 1453), and the biggest of all, Kaiser Wilhelm's Paris Gun.*

---

*At the end of World War I, military leaders on both sides were looking for a way to break the trench-warfare stalemate on the Western Front. The German general staff decided that a terror weapon, to be used against the city of Paris, might be the way to turn the tide of battle. In response to this idea, the German arms makers built the largest artillery piece in history, known to the Allies as the "Paris Gun."

In early 1918, the largest cannons extant could hurl a massive explosive shell about 23 miles, or the approximate distance of an Olympic marathon. In what takes a long distance runner about two and a half hours to cover, the big guns of World War I could hurl a 200-kilogram high-explosive shell in eighty seconds.

If they could find a way to bombard Paris, thought the German generals, the terror that would ensue could change the entire outcome of the war. The French government, it seemed logical to assume, would be heavily pressured by a suddenly vulnerable civilian populace and would be forced to negotiate for peace on terms favorable to Germany. But the German artillery emplacements were much farther than 23 miles from Paris, hence the need for an extra-long-range gun. The munitions makers set to work on designing this huge gun, the Kaiser Wilhelm Geschütz, or Kaiser Wilhelm's gun.

The Paris Gun looked great on blueprints, but it never worked nearly as well as the generals had hoped. There were seven barrels made but only two carriages, so effectively only two guns were ever in use at any time. Also, the massive guns developed such high internal pressures and temperatures that they wore out very quickly. Most important, at such great distances the accuracy of the Paris Gun was terrible. It could never come close to hitting a specific target and was of value only as a way to annoy and frighten French civilians. It eventually bombarded the suburbs of Paris from March to August 1918. The gun was positioned on movable railway cars located in the dense forests of Crécy, about 70 miles from the city.

In the end it had little effect on the war's outcome. The guns were withdrawn and dismantled in the face of the Allied advances in August 1918; one spare mounting was captured by American troops near Château-Thierry, but no gun was ever found by the Allies during or after the war.

The big air guns in the Millsboro soybean field have a lot in common with the Kaiser Wilhelm Geschütz. Both types of guns share similar parts: the breech, the barrel, and the mount. They differ in that the big air guns have pressure tanks instead of a combustion chamber, and a fast-acting valve instead of a percussion cap or fuse. But when any big gun goes off, whether it's the Paris Gun, the World War II battleship *Missouri*'s 16-inch deck guns, Jules Verne's Columbiad, or the Big 10-Inch, it is truly wondrous and terrible at the same moment, depending on whether the projectile is coming or going.

THE TECHNOLOGY OF LARGE-SCALE AIR CANNONRY

The technology behind making a big air cannon is not complicated—it's more a matter of scale than of intricacy. The barrel of an air cannon is typically a large-diameter steel or aluminum pipe. They are usually salvaged from industrial or agricultural applications, aluminum irrigation pipe and industrial iron and steel process piping being favorites. The pressures inside air guns can reach astounding levels at the moment the valve is opened, although they quickly lower to a more reasonable figure as the pumpkin slides up the barrel and the volume of gas inside the gun expands.

There has to be a way to place the ammunition, in this case a hard-shelled, 10-pound, gray-white Lumina or Casper hybrid pumpkin, into the cannon. The door and pumpkin-holding area inside the cannon are known in pumpkin-chunking circles as the breech. Often, this is nothing more than a bolt-on steel plate near the bottom of the barrel and a pipe placed crosswise on which to balance the pumpkin and keep it from sliding back into the valve opening. Sometimes the more intrepid pumpkin-gun designer will manufacture an elaborate hinged door with a hefty steel pumpkin-holding grate.

The barrel is mounted on a support structure, called a carriage. The carriage performs a number of functions in the operation of a big air gun. First and foremost, the carriage holds the gun in place while it is fired. The carriage also allows the contraption to be moved to its place on the firing line. Perhaps most important, a good carriage provides the ability to aim the pumpkin in terms of direction and altitude.

It's not hard to aim an air cannon. Most of the time, they are simply hauled into place by a truck and then fine-tuned by several large men lifting and tugging on the carriage until the general track of the barrel is in the right direction. But calculating and setting the angle of the barrel is a different story. In fact, the angle made by the barrel and the ground is a key factor in determining the distance and height the projectile travels.

According to the laws of physics, a projectile will travel the farthest distance when it leaves a barrel held at 45 degrees from the horizontal. But many air gunners, especially the T-LAR guys, just don't believe it. Perhaps considering the wind or the odd aerodynamic characteristics of a pumpkin in ballistic flight, most gunners set their firing angle lower than 45 degrees—closer to 40 degrees, with some going as low as 35 degrees. The proof is in the pudding, or in this case the distance between the barrel and the pumpkin splat.

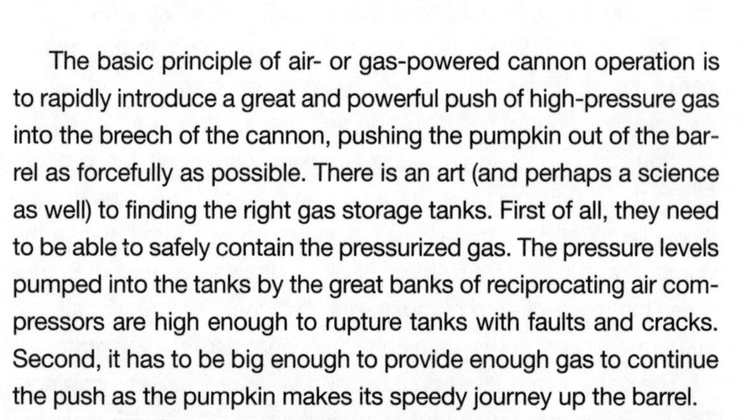

The basic principle of air- or gas-powered cannon operation is to rapidly introduce a great and powerful push of high-pressure gas into the breech of the cannon, pushing the pumpkin out of the barrel as forcefully as possible. There is an art (and perhaps a science as well) to finding the right gas storage tanks. First of all, they need to be able to safely contain the pressurized gas. The pressure levels pumped into the tanks by the great banks of reciprocating air compressors are high enough to rupture tanks with faults and cracks. Second, it has to be big enough to provide enough gas to continue the push as the pumpkin makes its speedy journey up the barrel.

But once the pumpkin is out of the barrel, the show is over as far as the gun is concerned. Any extra gas pressure simply vents to the outside and is wasted, so a gas tank that's too large is simply a waste.

The release of the high-pressure air is controlled by valves. A valve with a too-small opening, even if it opens very quickly, will retard the buildup of pressure inside the gun and hinder performance. A big valve that opens too slowly will do the same. The key is to find a big, fast-acting valve that opens the floodgate of high-pressure air or gas. Some builders use simple ball valves, while others go for more complex lever-operated compound valves.

In the end, though, all of this stuff—the barrel, the mounts, the valve and trigger—is secondary to the result, which is always the same: a teardrop-shaped mass of gooey pumpkin flesh in the middle of a Delaware farm field. <

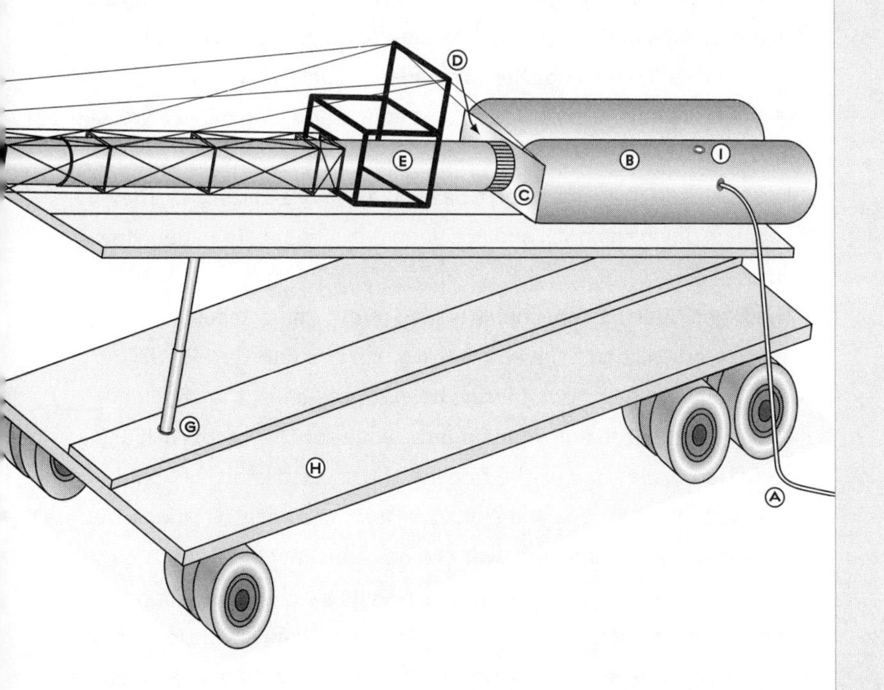

A High pressure hose connection to air compressor

B Air reservoir

C Fitting to connect cannon barrel to air reservoir

D Valve (valve must open quickly)

E Loading door (pumpkin is loaded through door; held in place by internal grating)

F Gun barrel (10-inch-diameter aluminum or steel pipe)

G Gun barrel support structure (angle iron)

H Trailer

I Emergency pressure release valve

> > >

Two hundred and eight miles to the north of Millsboro, the Central Park cops on horseback are still giving you the eye. To be sure, New York City cops have seen everything. Almost certainly they have already sized up the group—a dozen people in a circle, trying to look nonchalant, although they appear to be paying an inordinate amount of attention to a collection of white plastic plumbing supplies. Still, this is America. You are not here to cause trouble, and this is actually a legitimate scientific experiment of a sort. For a long minute you simply sit tight.

The mounted policemen don't pay you any more attention. They clip-clop on down the path and are soon out of sight. The countdown resumes.

The Right Guard has probably leaked out of the combustion chamber by now, so you don't figure there's much of a charge left. Still, it doesn't cost anything to try. Firing position resumed, the shoulder-held gun is presented and your fingers reach for the knob. With a quick motion, like a snap of the fingers, the knurled knob is flicked.

Ka-boom! With a bang and a muzzle flash, the potato rockets out of the gun barrel, arcing high over the oaks and maples. It lands away from people, in an open spot a hundred yards away—*splat.* Then all is still again. Heads turn. Your group is gathering some attention from those nearby. Several kids run up, eager to see what is going on. Some people look up quickly and nervously, but most just ignore you, going back to their reading or lying around.

A silly thing to do? Probably. Fun? Certainly. Fun enough to do again? Absolutely.

7. flame-THROWERS

It is very late at night and you are in a salvage yard piled high with scrap metal on the outskirts of Denver. There is a well-built and tough-looking man running directly toward you. Although you can't see his face because he is backlit by long tongues of petroleum-fueled flame, you presume that if you could see it, it would be a study in anger, a perfect portrait of unconcealed rage. For certain, you can see that his head is down and his arms are pumping wildly as he runs.

Two thoughts occupy your mind at this moment. The first is how much the silhouette of the man—the owner of the junkyard in which you are currently standing—resembles a pit bull.

The second is the ball of flame chasing the man running toward you. In the middle of that ball of flame is what used to be his tow truck.

"Well, guys," you say to your partners, "what the hell do we do now?"

> > >

Fire lights up the mind as well as the night. Fire is too physical, too tangible, and, contradictorily, too ephemeral a thing not to be extremely entertaining. It has endlessly fascinated people since the beginning of civilization. The entertaining qualities of fire started with rites and rituals around prehistoric campfires, moved to the medieval hearth, and continues in the fire rituals of more modern times.

Most ancient societies had a god of fire. The fire gods of the pagans demanded attention, supplication, and sometimes sacrifice. Although fire-worship ceremonies have mostly disappeared, vestigial performance and entertainment aspects remain.

Fire is ancient, but is it as old as the earth? No, it's not even close. It is surprising to many people to find out that fire is a relatively new phenomenon, geologically speaking. In fact, fire from combustion has been a part of nature only since the Devonian Age, or roughly 400 million years ago. Given that estimates of the earth's age generally run toward 4 billion years, fire, as we know it today, has only been present on the face of the earth for a small fraction of its total age. For 90 percent of the earth's existence, there was no fire.

For fire to exist, three components must be present. First there must be oxygen. Second, there must be fuel to burn. And finally, there must be a source of ignition, a hot enough concentration of energy or spark able to set off the self-perpetuating reaction of combustion.

Ignition has never been a problem. Even prior to the Proterozoic Era (a period during which multicelled plants and animals first evolved), there were plenty of sources of ignition—for instance, volcanic eruptions, friction from rock falling against rock, and lightning. But there was nothing to burn. Plants and organic matter to provide fuel for fires didn't appear until much later, and the level of oxygen in the atmosphere was wildly unstable.

In fact, until 150 million years ago, when the proportion of oxygen in the earth's atmosphere stabilized at 21 percent, there were long periods of time with either too much or too little oxygen to support what we know as "burning."

Eventually, conditions calmed and the earth became supportive of both people and fire. Although the dates are unknowable with certainty, it is clear that early humans began to manipulate fire early on. It changed how humans ate, allowing them to cook and therefore to eat things that were previously dangerous or even toxic. Some anthropologists believe that simply by learning to cook their food, evolving humanoids were freed from the evolutionary necessity of developing huge jaw muscles, which would have been imperative for chewing tough fibers. Forgoing the thick mandible-skull connections thereby allowed the skull to expand and the human brain to grow to its current size.

Fire shaped the social life of early humans. They bonded together in primitive communities for many reasons, not least of which was to communally tend a shared flame. The all-important fire needed to be stoked, tended, protected, moved, and most of all simply kept alive. New fire wasn't easy to find and even harder to make. The communal fire lit the dark of early humans' homes and allowed them to see in the night. It provided them with protection against bigger, more powerful predators. In short, human civilization has evolved with fire, and for that reason fire possesses an innate attraction to people.

The ancients found fire and its applications to be by turns entertaining and dangerous, enlightening and deadly. Control of fire meant control of society, pleasant and otherwise. It also implied that humans had gained some level of control over nature. The relationship between the two extreme natures of fire—fire as a weapon and fire as entertainment—is one that scholars and artists have recognized and explored.

Whether it really all happened exactly as described by the Roman historians is a matter of intense academic debate. But scholars agree that in the fourth century B.C., the Greeks dabbled in the use of fire against seaborne attacks. There is ample evidence that the early Greek city-states built siege defenses, including a sort of giant fire-belching bellows, designed to sweep away an attacking naval fleet with a fiery blast of ignited pitch and petroleum.

Several centuries later, the Chinese made extensive use of fire weaponry, including the weird yet cunning "fire ox," which was just what it sounds like. The Chinese would take an ox and then yell at it, poke it, and generally annoy it until it was really angry. Then some brave (or unlucky) soldier would go in and strap a big bucket of flaming goo to its rump. He'd give it a final slap to send it stampeding toward the enemy camp. Unfortunately, it was somewhat difficult to calculate the exact trajectory of an agitated ox with a bucket of fire strapped to its behind. Where the fire ox would wind up was affected by a host of unpredictable factors that no doubt limited its effectiveness as a weapon.

Oriental incendiary warfare from the Middle Ages was nothing if not extremely resourceful. Even more bizarre than the fire ox was the fact that Mongols often made use of the napalm-like quality of burning human fat, which they rendered from the bodies of their slain enemies. Loaded into barrels and set aflame, the fat was used as grisly (and probably gristly) ammunition for their catapults.

In general, the overarching idea of fire-based warfare is to spread terror by launching incendiary projectiles at the enemy, which over time led to the creation of the modern flamethrower. The earliest Levantine flamethrowers were simply hollow cylinders of wood filled with burning material, such as sulfur or liquefied pitch. These weapons worked like a big bellows-powered air gun: The tube was aimed at, say, an enemy ship, and a blast of air from the bellows would jet the flaming material toward its target.

The first fire projection device used in modern, mechanized warfare is usually ascribed to engineer Richard Fiedler. In 1901, he submitted drawings and models to the German army for a device he called a *Flammenwerfer* (which translates to "flamethrower").

Fielder submitted several designs, and two were made into prototypes. The effective ranges and squirt times (the maximum time duration of the fiery spray) vary according to accounts of the weapon but seem to have been in the area of 30 feet to over 150 feet, and 15 to 45 seconds, depending on the size of the flamethrower.

There was a small flamethrower called the *Kleinflammenwerfer,* which was capable of being carried by a single soldier, and a much larger one called the *Grossflammenwerfer.*

The *Kleinflammenwerfer* used compressed gas to carry a stream of ignited petroleum toward a target. The *Grossflammenwerfer* was apparently a defensive weapon, and was most suitable for use in trench warfare. It was to be useful in situations where opposing forces were dug in, at distances of less than 150 feet.

One of the most intriguing aspects of the concept to the German general staff were the expected psychological effects that a 20-yard-long spray of liquid fire would have on both sides in a battle.

The German army began using the device in 1911 by creating a group of specialists who were trained in the use of *Flammenwerferapparate.* The specially trained flamethrower-wielding infantrymen operated the devices by pressing a lever that caused gas to push the flammable liquid through a rubber tube and over an open flame inside the nozzle of the device. The flame ignited the pressurized fuel and sprayed in the general direction of where the operator pointed it.

The German flamethrower battalions were first deployed in 1915, the second year of World War I, when they were used in action against the French trenches at Verdun.

The weapon was impressive to be sure, but it did have serious deficiencies. It was bulky and difficult to maneuver. Further, it was hard to operate. In practical use, it could only be fired safely from a trench and had very limited range, so its main application was in areas where the opposing trenches were less than 20 yards distant, and that was an unusual situation on the Western Front.

Moreover, it was said to be very difficult to recruit soldiers into the *Flammenwerferapparate* units, for if the targets could survive the mostly psychological impact of the weapon, then the operators of the clumsy equipment were exceedingly vulnerable. And because of the nature of the weapon, the flamethrower operators were never taken prisoner.

The next real innovation in flamethrower technology came in World War II with the advent of the handheld flamethrower. This long, gun-type weapon had an attached fuel tank that soldiers carried on their back. The backpack contained three cylinder tanks. The two outside tanks held a flammable oil-based liquid fuel, with an ingredient pedigree dating back to Constantine's Greek fire. The middle tank held a flammable compressed gas such as butane. A mixture of fuel and gas was fed through a pressurizing system that directed it out through two hoses to the gun itself.

One hose was connected to the ignition system in the gun. The other hose was connected to the main fuel tanks. The compressed gas tank supplied enough pressure to squirt the fuel out through the gun. When the operator squeezed the trigger lever, it pulled the rod and the attached plug backward. With the valve open, the pressurized fuel flowed through the nozzle. A World War II flamethrower was powerful enough to shoot a fuel stream as far as 50 yards. Imagine standing on the goal line of a football field and lighting a cigar for a person standing on the 50-yard line.[1]

ENTERTAINING FIRE

In 1979, the performance art group Survival Research Laboratories performed its first show in San Francisco. SRL is a confederation of anarchic machine artists who specialize in building robotic,

THE TECHNOLOGY OF
LIQUID FUEL FLAMETHROWERS

The M1A1 shot a new and terrible type of liquid fire called napalm. Napalm was a mixture of gasoline and latex obtained from rubber trees. But since wartime rubber was in short supply, the formula was quickly reconstituted to use coconut oil. The M1A1 could shoot a stream of napalm more than fifty yards.

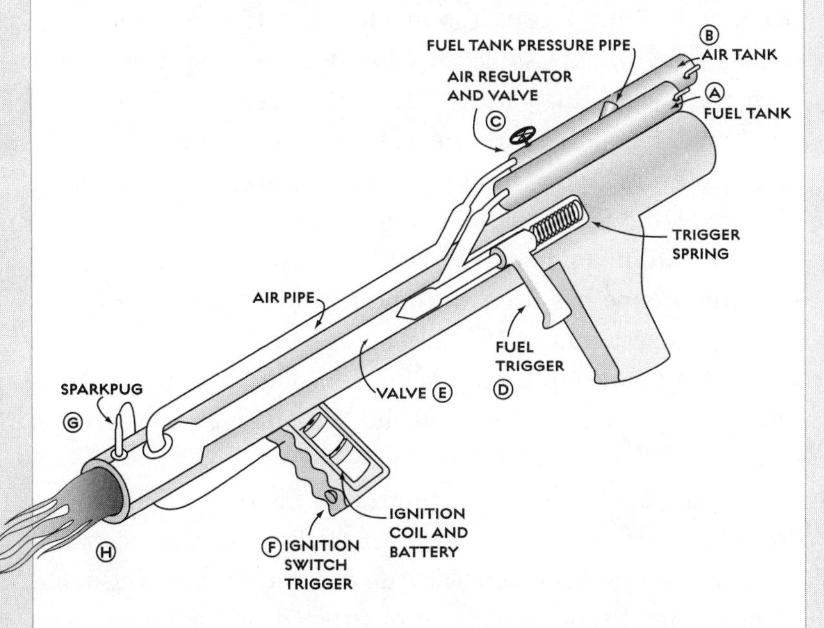

1. Fuel tank **(A)** filled with liquid fuel such as diesel fuel.

2. Air tank **(B)** is pressurized. This also pressurizes fuel tank.

3. Air valve **(C)** is opened; air leaves muzzle.

4. Operator pulls fuel trigger **(D)**. Fuel valve opens **(E)**. Fuel/air mixture is forced out muzzle.

5. Operator pushes ignition switch trigger **(F)**.

6. Spark plug energizes and ignites fuel/air mixtures **(G)**.

7. Ignited fuel/air mixture shoots from muzzle **(H)**.

remote-controlled machinery. SRL performances often involve violent imagery, loud, repetitive sounds, and very often fire.

There have been more than forty SRL performances, and typically they have involved fire shooting out from robots, roaring from the exhaust of jet engines, arcing from high-voltage discharge machines and, most of all, blasting from flamethrowers. Machines, not people, are the focus of SRL performances. Despite their importance, the machines have simple, benign-sounding names that belie their menace. For example, the Pitching Machine is much like a regular baseball pitching machine in form, except it is much larger. It is powered by an engine from an old Cadillac and flings 6-foot-long two-by-fours 25 percent faster than Roger Clemens's best fastball. There is the Air Blaster (a big air gun that shoots concrete projectiles). And there are many machines that involve fire—the Flame Hurricane, the Spark Shooter, and the V1.

Mark Pauline founded Survival Research Labs in 1978. Since then, the troupe has been an important incubator of ideas and techniques pertaining to the Technology Underground. Alumni of SRL have gone on to become key contributors to the communities of Burning Man, Tesla coil builders, and high-power rocket enthusiasts.

In 1982, Pauline lost four fingers on his right hand when a home-built rocket engine (remember the Estes Company's warning about backyard bombers) blew up while he was building it. His mangled hand was partially reconstructed by his doctor, who removed two toes from his foot and transferred them to his hand. It appears that this experience deepened his dedication to his art. Since then, Pauline has been the mentor to and inspiration for a large group of machine artists, including Kal Spelletich and the well-known robotic artist Christian Ristow.

Typical Ristow performance venues are places such as Burning Man, outdoor gatherings, and junkyards. At his performance's climax, one of the robots, an animal-like tracked bulldozer with a three-pronged hydraulic pincher called the Subjugator, incinerates a pile of wrecked machinery.

The two-and-a-half-ton, 16-foot-tall Subjugator is a robot of immense size and power. Ristow manufactured it by taking parts of different types of construction equipment and seeing what he could come up with. Its core came from the frame of a Bobcat-style front-loader, but it's been so heavily modified that its origins as a piece of construction equipment are all but unrecognizable. A 37-horsepower V-4 industrial engine is the Subjugator's heart.

The robot has a distinct personality, and it is an unfriendly one. The animated claw-like extension seems to have a will of its own, independent of the vehicle on which it is mounted. The pincers on the end of the arm are fang-like, biting off big hunks of its victims and sort of swallowing them up. Two hydraulic systems, deep inside the interior, power the custom-built three-fingered claw.

The whole contraption is basically a mechanical carnivore—a machine that eats other machines. It is the machine that sits atop the robot food chain.

Some fire-as-entertainment enthusiasts install flamethrowers at the back of their cars to release an impressive ball of fire when they stomp down on the accelerator. Entertainers often include flamethrowers as part of elaborate pyrotechnic displays.

In the simple but insanely dangerous practice of fire breathing, performers turn their own bodies into flamethrowers by taking a low-volatility but highly flammable fuel (such as paraffin oil or kerosene) into their mouth and spraying it out. The fuel is spit out through the flame of a torch that ignites it, creating an impressive human breath of flame.*

*"Don't try this at home!" That was the caution that Don Herbert gave every week on his Saturday-morning television program, called *Ask Mr. Wizard.* It ran from 1951 until the early 1960s. Still good advice.

Almost no street performer can gain attention, adulation, and hat money as quickly as a fire breather. There have been some fire breathers among the street performers in the touristy neighborhoods of San Francisco, such as the Embarcadero or Fisherman's Wharf. But there are other neighborhoods where the tourists don't often go—Hunter's Point, Tenderloin, Bayview, and Butchertown. These districts have significantly higher populations of transients and street people. Rents are lower. So this is where many of San Francisco's well-known community of radical self-expressive mechanical artists have their studios.*

No one would ever mistake Kal Spelletich's Butchertown neighborhood for North Beach or Nob Hill. His studio is located in an area of warehouses and decaying industrial buildings. Tourists have no reason to come here.

Spelletich lives in a loft above his studio, along with four or five other people and a six-toed dog. Each person has his or her own separate cubicle built from two-by-fours and plywood. He is the landlord, of a sort, and the others in the studio sublet space from

---

*There is one interesting area south of downtown nestled up against San Francisco Bay, close to the very non-trendy, very non-touristy industrial shipyards of eastern San Francisco. This area is called Butchertown, because a hundred years ago this was the meat district. Newspaper and magazine accounts of the time describe Butchertown as a cesspool, a place where animal offal left over from the slaughtering process was simply pushed into the surrounding swamps to rot, where blood and discarded guts were sluiced directly into San Francisco Bay. It was a place where the resident meat cutters, called the Butchertown Boys, had to contend with great buzzing clouds of flies and huge swarms of aggressive rats. When the wind blew in from the bay, the Butchertown stench gagged office workers downtown, wafting as far as Nob Hill.

Butchertown did not take a backseat to Upton Sinclair's jungle in terms of worker misery or danger to the public welfare. For years, people complained about the smell and filth, but little was done. Then in 1907 the miserable conditions caused by so many rats and other vermin led to an outbreak of bubonic plague in the area. Apparently, nothing motivates a city to bold and decisive action like a bubonic plague outbreak.

The city's leadership tried to clean up Butchertown. Despite some decisive action taken by the city fathers, including inspections, condemnations of the worst offenders, and tighter regulations, it took a very long while to clean up. In fact, sanitation problems related to meat cutting persisted until the last abattoir in the city was closed in the 1970s.

him. The residents form an oasis of creativity and liveliness within Butchertown's mostly bleak confines. Butchertown is host to an eclectic assortment of workshops and studios, none more interesting than Kal's studio. It is a post-apocalyptic wonderland of metal, fire, iron chips, and oil; a gritty, grimy, industrial place, crammed with machine tools and workbenches, welders, and robots.

It is in Spelletich's studio that the overall relationship between fire, art, and danger begins to come into focus. Some, maybe most, of this stuff has a less-than-pleasant genesis. How can something as deadly and malevolent as a military flamethrower become a cool, neat, and worthwhile thing? Through the work of people such as the machine artists of San Francisco, the linkage between the military and entertainment uses of fire is being defined.

Spelletich is one of several visual artists in the Bay Area who specialize in media that include dangerous-looking machines and fire. He is a tall, youthful-looking midwestern transplant with a big toothy smile. He occasionally travels to Mexico to participate in festivals and celebrations that involve copious amounts of fireworks. Although Spelletich has no obvious missing limbs or visible cranial mending plates, it was apparent from the ugly burn scar on his hand that occasionally Kal gets a little too involved for his own good.

Kal pointed to his hand. "Have you ever seen a Mexican fireworks display? A real big one, like they do in those towns known for their tradition of building fireworks? They're incredible. Dangerous, but incredible. I drove down there for a big festival. It was in the town square, the *zócalo*. Tons of fireworks, shooting off from everywhere. I knew it would be wild, so I came prepared for it. I put on my motorcycle helmet and was pretty much encased in leather. Leather coat, pants, boots, gloves—you know, a biker-style suit of armor.

"I wanted to get into it, to experience it as fully as possible, so I walked down to the front, as close as I could to where the fireworks were being ignited. They were going off everywhere—rockets, aerial bombs, phosphorus showers, mortars, towers of flame. The Mexicans build very high fireworks display platforms, and they absolutely load them up with wheels and wheels of rockets. They spin around and throw off showers of fireworks. It was great.

"That is, it *was* great until one of the rockets came right at me and somehow got wedged in tight between my coat and glove. It traveled up my wrist toward my hand and got stuck there." He pointed to the 3-inch white scar on his hand.

He grimaced. "Hurt a bunch when it happened. It's still a bit tender now and then."

Spelletich is a man very familiar with the entertainment value of fire, and he is the motive force behind the machine art group SEEMEN. He is a philosophical yet outgoing person who builds a lot of mechanical stuff. Spelletich works in many edgy types of entertainment technology, but flamethrowers are his passion. His studio is filled with flamethrowers as well as various types of robotic art and articulated mechanical technology upon which the flamethrowers are placed. In fact, a primary thematic element used in SEEMEN's work appears to be the primal and joyful use of fire.

There are great numbers of spent fuel canisters strewn around the place, and LP gas tanks all around—on the floor, out back, on tables, mounted on artwork, held in place by twisted wire and band clamps. Spelletich has, through long experience, become very knowledgeable regarding the best ways to control the size and quality of the flames. He uses ball valves, regulators, solenoid valves, integrated circuits, and all sorts of strange and arcane control gear to provide a host of big-combustion and tension-filled experiences for his audience.

Where does Spelletich's deep and abiding interest in things such as flamethrowers come from? He says he likes them because they are antidotes to the ordinary and the vicarious. That's why his exhibitions and performances are well attended. He says that "the more people stay at home sitting in front of computers, playing video games and watching movies on video, the more that they will be drawn to events that allow them to witness and experience their own mortality and humanity."

Besides the ongoing flamethrower art installations, Spelletich's other area of intense specialization is something he calls "biomorphism." He explains it as a refined type of human-machine interface, one where a mental connection or link between the machine and the person who is encountering it is created.

For example, he's built a contraption of metal tubes that surround a seated participant. In very close proximity to the participant, columns of flame issue from various points on the framework; the length and intensity of the flames are controlled by a solenoid valve, which in turn is controlled by a small integrated circuit board. The IC board came from a surplused hospital EKG machine, and the circuitry is designed so that the more stress the participant feels, the higher and hotter the flames become. This sets up a feedback loop: As the person in the chair becomes more nervous, the flames become higher. The machine and human become a single system, each controlling and being controlled by the other.

"I do want to give people new and unique experiences through my art. Really, the participant and the machine combine to form the overall artistic value. Biomorphics is more than just a way to connect the audience with the art. It actually fuses them together to make the art."

The SEEMEN troupe's use of fire and flame art extends to most of the machines and artistic installations they build. There is the Levitator, a machine in which the participant stands and rocks backward and forward, at some points almost levitating, while flames shoot out from a metal crown near the person's head. There is the Lie Detector Halo, another combination of flamethrower and biomorphic feedback device, in which the flames are controlled by a voice-stress-analyzer chip.

Typically, a SEEMEN exhibition takes place in a large hall or studio, somewhat clandestinely, and is often advertised only by word of mouth. Handbills and advertisements would attract too much unwanted attention from city fire departments and insurance companies. So one person tells another person, who tells another, who sends an e-mail to several others; often a hundred or more people will converge to experience the SEEMEN performance.

At a recent Los Angeles show, a few new items premiered, ones that Kal has spent considerable time developing, along with some older pieces. The Fireshower is a relatively recent innovation. It is a large tube-shaped metal framework of piping and nozzles. During a SEEMEN performance, an audience member stands within the tube while flames swirl up, down, and around him—he's literally taking a shower of flame. The fire never actually touches the user, but the effect is novel, frightening, and even breathtaking.

Joe Riche is a machine artist and a man who takes great interest in the entertainment value and artistic symbolism in fire. Joe is a descendant of the French Canadians who first settled the bayous of the Mississippi delta. He hails originally from Metairie, a town near New Orleans, and attended a state college in Lafayette, the capital of Cajun country. As a young man, he became acquainted with the junkyards and metal scrap heaps of southern Louisiana, and he found these were special places.

The salvage yards that bought and sold surplus parts and metals from the Gulf's offshore oil industry always contained mountains of terrific cast-off stuff. In fact, Joe makes a compelling case that no place in the country can compare to the Louisiana–East Texas coast in terms of the variety, quality, and price of the scrap metal and surplus mechanical equipment available. A tinkerer's mega-sized supermarket, this area of the Gulf Coast is a sort of metallic paradise of scrap yards heaped high with used drill bits, broken compressors, old sucker rods and pumpjacks, well casings, hydraulic cylinders, filter housings, and other goods worn out through a hard life of service in a narrow hole a couple of miles underground.

In these scrap yards, Joe Riche finds a wealth of raw materials. Even more important, in the twisted metal and dented air tanks he finds the basis for his own brand of artistry and self-expression through technology. He takes a lot of the stuff he finds here back to his workshop, where he honed a knack for extreme tinkering, metal sculpture, and machine art.

In the mid-1990s he left the Louisiana bayous for higher education in the mountains of Colorado and stuck around there. Now he works on a variety of artistic projects as well as the more prosaic architectural and furniture designs that pay the rent. But his passion is still avant-garde mechanical projects, the most important currently being the Motoman Project. Motoman is an artistic collaboration between Joe and his partners, Eric and Zac. Motoman is moving, kinetic, high-powered art. It's a collection of radio-controlled and heavily modified earth-moving equipment, robotic arms, propane-fired jet propulsion units, and of course flamethrowers, all carefully woven and orchestrated into an artistic performance.

Riche has thought about the idea of self-expression through technology. "Basically I cannot see doing anything else," he says. "My partners and I are driven as artists to work in this genre. Working within the gallery system is nowhere near as adventurous as putting diesel-spewing flamethrowers, radio-controlled machines, high-voltage equipment, and roaring pulsejets in the path of our audience. We give the audience an opportunity to experience true reality.

"Everything we use at a performance is fully operational and dangerous but under controlled circumstances," he explains. "I think the Motoman Project counteracts the synthetic bubble that most people live in."

Obviously, it's a difficult type of art to explain. It's even more difficult to obtain serious recognition working in the genre. The hours are too long, the pay is nonexistent, and most of the audience understands, at best, just a fraction of what you're trying to say. To top it all off, finding a venue willing to put up with the noise, the grime, the oil stains, and the shrapnel makes it tough to find a gig. So most machine artists will play wherever and whenever they can find a suitable space and a host with a tolerance for lawsuits.

On one particular night, they were scheduled to work in front of a large and well-chosen crowd, a group likely to sympathize with their artistic politics. Their live performance with a high-powered, liquid-fueled, fire-shooting cannon was set to take place at a show put on by a troupe of performance artists with whom they have a passing acquaintance. The troupe, called the Know Nothing Family Zirkus Zideshow, also hails from New Orleans.

The Know Nothing Family Zirkus Zideshow bills its show as "The Greatest Evening You Will Ever Spend in Hell." This night, the act showcases the unusual talents of three performers, Dr. Eric von Know Nothing, aka the Enigexclamation; Micki Luv, the Runaway Rockstar; and Flag Blashpoint, the Mad Monk. Their show runs a wide gamut that includes fire baton twirling and fire eating; escape artistry; walking, crawling, and then actually slithering over broken glass; and sword tricks that involve using a machete-like blade to completely cleave a watermelon placed on the stomach of a near-naked performer.

The Zirkus's manager lined up a gig at an automotive recycling center (otherwise known as a junkyard) on the outskirts of Denver. When Zirkus invited Motoman to bring its stuff on over to the junkyard for a performance, the members jumped at the chance. The invitation to perform was proffered because the Zirkus heard about Motoman's new high-performance fire cannon and thought it would be just the thing to keep the audience interested and involved during the intermission of the Zirkus show. Although there doesn't necessarily seem to be an obvious artistic synergy between the Motoman Project and Zirkus Zideshow, the guys took it because, well, a gig is a gig.

> > >

You and the members of the Motoman Project have arrived at the evening's venue and have begun to set up the newest piece of equipment, a liquid-fuel flamethrower capable of pumping out 50 feet of high-intensity diesel flames with each shot.

Now, it likely comes as no surprise that junkyard owners aren't generally known for their support of the arts. And when you and the team from the Motoman Project roll up to the junkyard and start assembling the flamethrower, there is a problem. The yard owner (a guy who, as one of the Motoman crew puts it, is "a stereotypical junkyard owner, full of grease, and a guy who takes no shit") takes one look at the cannon and calls you over, shaking his head. "No way, guys. Not here."

It takes a while to get the green light. You and the others apply some serious persuasiveness and charm in order to get the owner to allow you to use the cannon. You paint a vivid verbal description of the artistry of the flamethrower. You explain the philosophical statement within it, the symbolism, the visual poetry involved, the Dadaist underpinnings. After a fair bit of convincing (and plying with beer), the owner finally agrees. But, and this is a big but, he gives you one important warning.

"Look," he says. "See that truck there?" Grim-faced and deadly serious, he points to his tow truck and tells you clearly and unequivocally that while he doesn't care much about you damaging almost anything here—this is a junkyard, after all, and most of this stuff is harder and tougher than you are—you'd better not fuck up his truck.

So that's it. There is really only that one requirement—just don't let anything happen to the tow truck, and everything will be fine. No problem.

Zirkus takes its break for intermission, and it's time for the liquid-fuel cannon to show its stuff. You load it up with fuel and then, to the delight of the assembled crowd, it lets go with a terrific blast of flame. But for some reason, instead of blowing forward as expected, the huge ball of liquid fire shoots straight up into the air before falling right back to earth, and in the process it torches one of the wrecked cars in the lot. The junkyard owner does not look pleased. But the crowd is thrilled and roars its approval. "More!" they holler. "More fire! More flames! More everything!"

The pyrotechnics, added to the freakish sensory experience of Zirkus, light the crowd's emotional fuse. The demand for more excitement, more energy, is palpable and can be slaked only by more of the same—the unregulated and unthrottled emission of several thousand BTUs of pure combustive energy.

> > >

Months later, after time for reflection, Joe deconstructs the evening and breaks down the action.

"So we did some adjustments after the first flamethrower shot. The kerosene didn't go where it was supposed to at all, but since we did not really know what the problem was from the first, we were kinda lost. So here came the time for the second shot. It goes off, but at that exact second, the wind changes, and gallons of flaming diesel land right on—as everybody probably knew it would—the tow truck.

" '*Oh, shit!*' I yelled. I knew that the junkyard owner wasn't gonna let me go without taking some of my teeth with him." Joe pauses for a second, eyes wide, remembering the scene vividly.

"The situation is this: The junkyard owner is running toward us at top speed. He is sort of a big guy, his head is down, and he is coming at us fast."

They steeled themselves for whatever would come next.

"But then he looks up and I see his face," says Joe. "And—get this—he's smiling. He's actually smiling and laughing. I guess when he saw the size of that blast of flame he forgot about the truck. Maybe he was too impressed or maybe he was drunk, but he wasn't mad. In fact, he loved it."

The Motoman crew extinguished the flames and then checked out the truck. Incredibly, it was still perfect. Despite the fact that great gobs of burning petroleum had landed squarely on it and burned furiously, there was no damage, no burned paint, no charred Zideshow performer residue, no nothing. It seemed almost a miracle. Still, timing is everything, so they "packed up the flame cannon and hauled ass out of there before he had a chance to sober up."

8. electrostatic MACHINES

You can feel the hair on your head stand on end, not from fear but from the effects of the electric field that fills the room. Here, in the darkened auditorium at the Theater of Electricity at Boston's Museum of Science, there is a pair of 50-foot-high electrodes that look like gigantic silver golf balls perched on railway-trestle-sized tees. This powerful-looking machine, four stories tall and about half as wide, generates an electrical potential of somewhere in the neighborhood of 5 million volts across its 23-foot-wide electrodes. It seems much too big, and much too expensive, to have been built simply as a demonstration for a science museum.

But it is here, right in front of you, just across the theater, and so near at hand. It is wonderful and terrifying to be so close to an apparatus of such power and seeming danger.

You've come to the auditorium early in order to land a good seat for the premiere performance of Zap! Music for Van de Graaff Generator, Robots, Instruments, and Voices. When the performance begins, the conductor, who is also the piece's composer, rises in a rounded steel cage on a hydraulic pole and is silhouetted by the lights from the Van de

Graaff's rounded electrodes. When energized, the electrodes become frenzied, and fiery, coruscating sparks—a dozen feet long—leap with deep, resonant cracks from the electrode's steel sphere. The lighting encircles the cage but does not enter it. The conductor is unharmed, protected by an electrical phenomenon called the skin effect.*

Outside the cage, away from the arcs, a group of musical performers work with the noises produced by the Van de Graaff to synthesize a very unusual musical composition. The woman inside the Faraday cage is Christine Southworth, a young MIT graduate with degrees in mathematics and music. Her quest to combine both disciplines resulted in a high-voltage stage performance that is electrifying (literally and figuratively). Standing in the Faraday cage just in front of the giant Van de Graaff at the Museum of Science in Boston, she's devised an electronic music concert combining a flute, electronic keyboards, a cello, a guitar, various types of percussion, and the Van de Graaff itself.

Aside from amplifying the guitars and drums, electricity is integrated into the overall program as buzzes, hums, rings, drones, and cracks. It is not neccesarily hummable music, but it is certainly unique. The musicians work hard to keep the beat going, even when the electrical performers are arcing in a different rhythm from what the musicians and their robotic co-instrumentalists are playing.

Southworth is part of an avant-garde artistic collaborative called Ensemble Robot. The ensemble's goal is to make music, both simple and complex, producing unusual patterns of sound from a variety of sources—strings, pipes, drums, wooden keys, and occasionally a Van de Graaff generator.

---

*Electricity is a lazy entity and will always choose the path of least resistance. So it'll travel over the surface of the cage, which creates a closed circuit, rather than through it. But if the conductor sticks any part of her body outside of the cage while the electricity is on, then that hand, finger, foot, elbow, or whatever will be considered a part of the circuit. And that would be a very bad thing.

>>>

As far back as Ben Franklin and his electric kite, people have always found the notion of pure electricity entertaining. Around 1780, simply capturing enough electricity in a glass jar to make a small spark on command was enough to excite entire towns to enthusiastic appreciation.

Throughout the eighteenth century electricity was a major entertainment draw. Electrical demonstrations took the place of theater. Electrocution of small animals, particularly birds, became a morbid but popular form of entertainment. The ability to supply a painful but momentary surprise shock to unsuspecting friends at parties passed for great wit and charm. And the general population began to fill lecture halls in hopes of witnessing the production of artificial electrical sparks. Independent lecturers, ranging from serious, learned academics on down to simple hucksters, traveled America and Europe, attracting large crowds willing to pay big money.

At the top of the itinerant "electrifier" (as they were sometimes called) pecking order were those associated with learned societies and universities in Stockholm, Paris, and Munich. High in status, possessing the ability to give cogent lectures, and traveling with excellent demonstration materials, these electrifiers commanded very high fees. Next on the status continuum were demonstrators and lecturers who were members of learned societies but who worked independently of their institutions. Next came the entrepreneurial types. These men often rented rooms and performed in public inns and taverns. Finally, there were the simple hucksters, trying to make a buck. They attempted to earn their living by hawking electrical curiosities, busking on the streets of large cities, and making sparks and doing electrically based tricks at fairs and marketplaces.

In the pre-television, pre-motion-picture days of eighteenth-century Europe, traveling shows filled a need for entertainment-starved townspeople. Of all the types of entertainment available, none had as much appeal as the "Physicks" shows featuring the wonders of electrical phenomena. For example, archives from Rotterdam in 1780 show that the cost of admittance to a wild animal show was two stuivers. A show with dwarves was twice as much, four stuivers. And a show featuring automata (robot-like machines that simulated nature and animals) cost about twenty stuivers. The electrical lectures were the most expensive ticket of all, costing as much as thirty stuivers for admittance.

Over time people became more sophisticated in their dealings with electricity, and small sparks would no longer suffice to enthuse a crowd. Bigger and more powerful electrostatic machines were designed, and a whole mechanical bestiary of hand-powered electrical devices was developed.

In the early nineteenth century, James Wimshurst, a British naval engineer, popularized a very advanced induction-static machine. The Wimshurst Influence Machine, as it was known, was the most famous of all the static machines.* Besides being relatively easy to build and highly entertaining (for its time), it became an important laboratory tool that produced the high voltage needed for the experimentation with high-voltage vacuum tubes that was taking place.

But it wasn't just Wimshurst machines that were being built. At the turn of the nineteenth century, there were more types of static-electricity devices and inductive generators than an electrifier could shake an electrode at. Electrical experimentation was a popular pastime, similar to the way computers entertain people today.

There were many variants on the theme of electrostatic generators. Besides the Wimshurst, there were the Ramsden Friction

---

*Electrostatic machines are electromechanical devices that produce high-voltage static electricity. The first machines made discrete sparks; later advancements allowed the machines to produce continuous arcs of electricity between electrodes. Such machines were fundamental to early studies into the nature of electricity. The first machines were called friction machines and were used in Ben Franklin's time as a research tool. Static machines were not particularly powerful or efficient and were supplanted by influence machines in the nineteenth century.

Machine, the Lebiez machine, the Voss machine, Lord Kelvin's Replenisher, the Toepler-Holtz machine, and the Dirod. The variations on the idea of a spark-making electrostatic machine were seemingly endless, reflecting the appetite for experimentation and scientific amusement of the time. There were Bonetti devices, Leyser machines, Bohnenberger's apparatus, Nicholson's doublers, and Wehrsen machines. And all of these basic designs lent themselves to variances and experimentation in quest of bigger, longer, and louder sparks.

The biggest, brightest, angriest, and most fulgent electrical discharges don't come from Wimshurst-style electrostatic machines. They don't come from quarter shrinkers or Tesla coils either. They come from an auditorium-sized pair of tall cylindrical machines. They are called Van de Graaff generators, and they produce cascades of sparks, electrical effluvia, and strong electric fields, which overflow and tumble down in long bright tangles from huge hollow metal electrodes.

Dr. Robert Van de Graaff built and designed this contraption at MIT back in the 1930s. The two large spherical steel electrodes are hollow, and there is space inside in which several scientists could sit and obtain an inside-out view of the man-made lightning. The electrode/laboratories are mounted on thick 25-foot-high columns of thermoset plastic called Textolite, a type of plastic resin similar to what old-time ashtrays and 1930s-era table radios were made from. Its insulating capabilities are second to none.

Milton Stanley Livingston, a professor at MIT, the director of accelerator projects at Brookhaven National Laboratory, and the inventor of the cyclotron, spoke glowingly of the Van de Graaff generator in his book on the subject, *Particle Accelerators*. In 1962, Livingston wrote this of the large Van de Graaff generator now in the Boston museum: "The popular appeal of such a gigantic generator has been tremendous. It is an awe-inspiring

THE TECHNOLOGY OF

VAN DE GRAAFF GENERATORS

The electromechanical foundation of a Van de Graaff generator is straightforward. Its main component is a moving vertical belt made out of an insulating material such as rubberized silk. A charge collector, called a comb because it looks like a hair comb, consists of an array of pointed metal needles connected to a solid electrical ground. The comb "sprays" an electric charge onto the lower end of the belt by a process called corona discharge. The continuously moving belt transfers the electric charge to the inside of a hollow metal sphere at the top of an insulated column, where another comb that is directly connected to the dome removes the charge through the same corona discharge process used below.

As the belt continues to turn, more and more charge builds up, causing ever-increasing electrical potential. Simultaneously with the buildup of the charge, the charge also spreads over the outer surface of the dome. The voltage difference between the charged dome and the ground grows and grows until the voltage becomes so high that the air surrounding the dome, normally a great insulator, breaks down and *zap!*—lightning occurs. A bolt shoots from the dome to whatever is in the path of least resistance. <

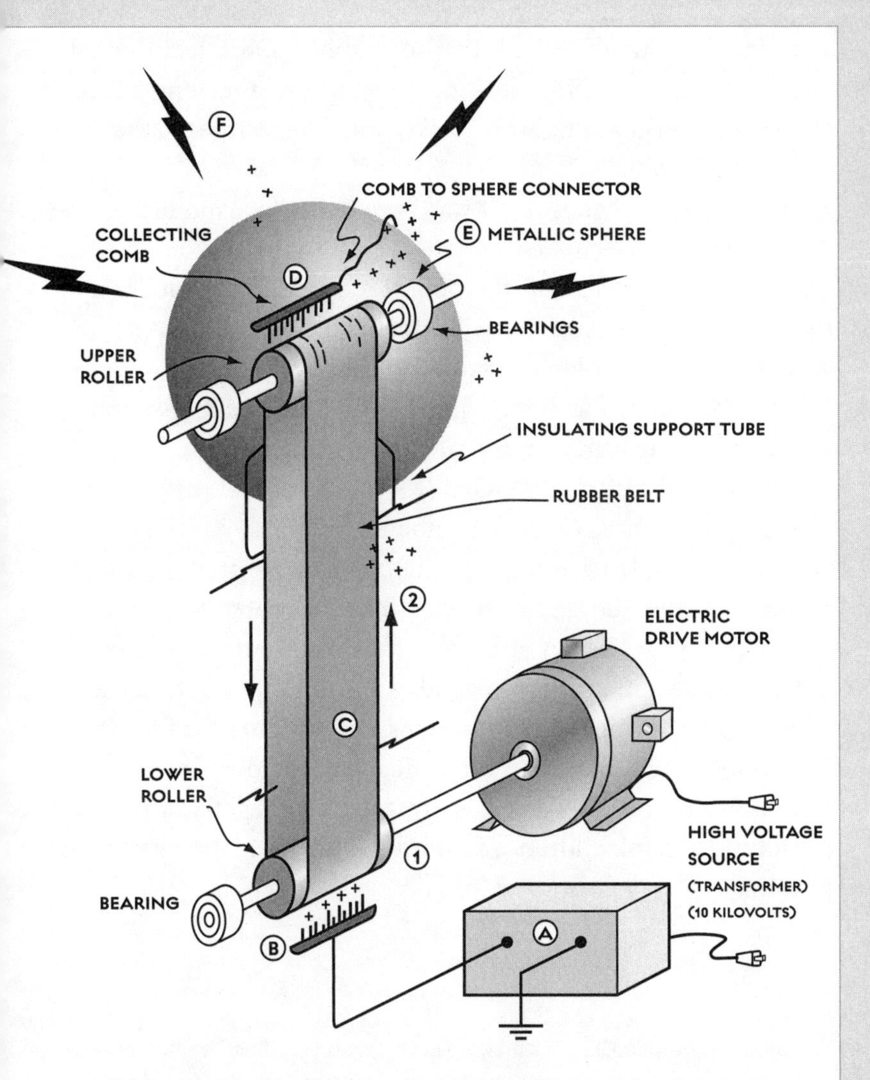

1. High-voltage source **(A)** "sprays" charge from comb **(B)** to belt **(C)**.

2. Motor turns belt, gets coated with ions.

3. Belt moves charge from bottom roller and comb to top.

4. The upper comb collector **(D)** catches the charge and moves them to the sphere surface **(E)**.

5. The longer the belt runs, the more charge collects on the sphere.

6. The charge difference becomes so great that electric discharges spontaneously occur **(F)**.

experience to stand beneath the huge spheres and feel the hair rise as potential increased, and then to see the long jagged strokes of man-made lightning as terminal discharges to the roof or down the column."

According to Livingston, it was even better to stand inside the 15-foot-diameter electrode when the power was turned on. The scientists would ascend into the hollow electrode via a ladder and then look out through the viewing port as the surface was charged with huge sparks striking just a few inches away.

Formally known as the Round Hill Van de Graaff Generator, this is one of the world's most famous machines.

The Round Hill Van de Graaff was one of the early atomic research high-energy electron shooters and a key piece of test equipment that led to success in the race for the atom bomb fifteen years after the generator was initially brought online. Its original purpose, before it became one of the most exciting, if didactic, museum displays around, was to accelerate atomic particles for use in high-energy physics experiments for the lab at MIT. After a life of service on United States government projects, it continues to work even into its septuagenarian years, delighting schoolchildren and adults with its giant display of physics.

>>>

The gigantic Van de Graaff machine is front and center, the human performers are crammed onto a small part of the first balcony, and Southworth is nearly rubbing up against the sizzling electrodes, protected by the birdcage-like shield. The music itself isn't bad, described by a Boston music critic as "a sort of minimalist concert music that vacillated between moody ambience and jazzy grooving."

Beyond the sound, the show entertains you on an instinctual or even neural level—a large, impressive, and inspiring display of the excitement that technology can offer.

Nearly eighty years after the generator was first built, the electromechanical descendents of Robert Van de Graaff's high-energy electrical device continue to play an important role in the fields of astronomy, high-energy physics, and medical research. But for those who experience it here, it doesn't need to justify its existence based on the argument of utility or service to researchers. It's appealing enough and impressive enough that you will long remember it even if it did nothing beyond delight visitors to the museum. This is pure technology that entertains. It's hard not to be impressed.

9. rail and coil GUNS

It's getting close to 1:30 A.M. Tomorrow is a workday. But your fingers continue to fly across the keyboard because there is a lot going on tonight. There is a critical mass of magnetic gun builders online, and there are several different threads and discussions happening all at once.

A forum member has posted a grainy Quicktime movie of a heretofore unknown and homebuilt magnetic weapon firing a projectile that crashes with loads of force and momentum into a sandbagged backstop. Besides the movie, someone finally got around to uploading several pictures of a magnetic linear accelerator whose existence was quite uncertain.

"Wow," you type. "The photo you posted of that kid's rail gun looks terrific! He says it attains a slug velocity of 1.4 kilometers per second? That would be fantastic! I want to see his schematic. . . . What's the web address?"

A while later comes a response: "Don't get too excited yet. That sounds like another grossly exaggerated, untested, bogus claim based on bad physics. He says his cap bank is about 6 kJ. So, a theoretical

1.3 gram slug, traveling at 1,400 m/s, will have an energy content of about 1,274 J or an overall efficiency of about 21 percent! Given that that's ten times as efficient as anybody else's gun, well, IMHO that's highly unlikely."

Another user has a different opinion: "Looks like the real McCoy to me. Check out the scans of the meter readings."

Once in a while, somebody pops up out of nowhere who really seems to know something about magnetic guns and gauss rifles—someone who has built something substantial and has posted movies and pictures to back it up. Tonight is one of those times.

> > >

Many first heard the term "rail gun" in the 1996 Schwarzenegger action movie *Eraser.* In the movie, the muscular hero is assigned to protect a computer programmer who agrees to testify against her boss, an evil defense contractor who took billions in tax dollars to develop rail gun technology and then attempted to sell the resulting weapon to terrorists.

There are nearly always significant differences between real technology and Hollywood technology. Both movie and actual rail guns use electromagnetic forces to fire projectiles. Real rail guns shoot projectiles pretty darn fast, but not as fast as movie rail guns, which are handheld devices that apparently fire at speed-of-light velocities and are capable of taking down battleships. The recoil from an 186,000-mile-per-second projectile-firing rail gun would be hard for anyone to absorb, even Arnold Schwarzenegger.

Hollywood imagination notwithstanding, rail guns and their close cousins, coil guns, are fairly popular in the world of extreme technology projects, both with the Department of Defense as well as with amateur radical tinkerers. Governments develop them in multi-million-dollar programs in top-secret national laboratories; big electronics component manufacturing companies develop them, too, because they have commercial applications; and talented amateur experimenters develop them just for fun.

The idea of a rail gun has been around for a while; it appeared in sci-fi novels predating *Eraser* by at least a couple of decades. But fictional works aside, in the mid-1980s Ronald Reagan proposed a missile defense shield for the United States, and soon quite a bit of research money was allocated for a host of advanced armaments, all of which were grouped into what was formally called the Strategic Defense Initiative (SDI) but popularly referred to as "Star Wars." At that time, work started in earnest on the development of technology that would use large, high-voltage energy discharges to magnetically hurl projectiles through the air toward incoming rockets. The theoretical muzzle velocities and accuracy of electromagnetic weapons were beyond just stupendous—given a large enough capacitor bank, an ultra-fast-moving blanket of protective shrapnel could intercept and shoot down anything the Soviets cared to send our way.

Twenty years later, with the Cold War won and the threat from spaceborne missile attack substantially reduced, the Department of Defense chose not to continue fast-track development for this type of weaponry. But still, the idea of electromagnetic weapons did take firm root in the minds of some extreme home experimenters. A host of different types of high-energy gun powered by magnetic forces have been successfully modeled and built by experimenters: coil guns, mass drivers, Gauss rifles, pulsed-energy weapons, and rail guns.

All of these use electromagnetic energy to accelerate a projectile and can be described as "linear motors." A linear motor is an electric motor that has had its stationary windings "unrolled" so that instead of producing torque and therefore rotational motion, it produces a linear force and motion along its length.

Linear motors fall into two major categories, low-acceleration and high-acceleration. Low-acceleration linear motors are used in applications such as magnetically levitated trains. High-acceleration linear motors are designed to accelerate a projectile up to a very high speed and then let it fly. They are usually used for studies and simulations of ultra-high-velocity collisions, as weapons, or as launchers for spacecraft.

While there are many minor variations, there are two main branches on the design tree of high-velocity linear motors—rail guns and coil guns. They are related in that they both use magnetic forces acting on a projectile to accelerate it down a gun barrel. But they differ in the way the magnetic force is applied.

In a coil gun, a series of electromagnets is placed at intervals radially, up and down the long barrel of a gun. The magnets are energized one at a time and in sequential order. The first magnet pulls the projectile toward it, and just before the projectile reaches it, the first magnet switches off and a second magnet slightly farther down the barrel turns on. The projectile continues farther down the barrel, pulled forward by the second electromagnet until it too switches off and a third magnet, again just a bit further down the barrel, takes over. The process continues all the way down the barrel until the projectile shoots out of the gun toward its final target; given enough magnets and a long enough barrel, the slug velocity approaches hyperspeed.

Building the coil gun's array of precisely timed and switched electromagnets requires complex circuitry. The time intervals at which the magnets are switched on and off are exceedingly short, and even small discrepancies in timing can ruin gun performance completely. Nonetheless, there are many coil guns out there in the workshops of amateur inventors. Most are strictly for demonstration purposes, using small, disposable capacitors and rather crude timing sequences. But they do work, and they make the point that building a linear electromagnetic accelerator such as a coil gun is within reach of almost any experimenter with enough desire to build one.

Rail guns may be the most impressive type of electromagnetic linear accelerator. In fact, some of the fastest muzzle velocities ever attained for projectiles larger than atomic particles were rail-gun-powered.*

The rail gun's ability to propel objects at speeds that are simply impossible for conventional (combustion or pneumatic) guns makes it an extremely interesting device to a lot of people. The most obvious application is national defense, as seen in the SDI development of the 1980s and 1990s. But there are many other, more peaceful applications as well, such as building a space launcher that would deliver payloads into orbit at a fraction of the cost of a rocket launch. Some researchers say that rail guns are just the thing for injecting fuel pellets into nuclear fusion reactors. There are also ideas for using rail gun technology in less flashy but highly practical metallurgical applications.

---

*The highest rail gun muzzle velocity ever recorded occurred during a test at the Green Farm Test Facility (aka the Electric Gun Research and Development Facility) in San Diego in the mid-1990s. The rail gun muzzle energy readings topped out at 8.6 MJ and the highest muzzle velocity recorded was around 9,600 mph—about Mach 12. By comparison, a high-powered rifle's bullet travels at less than 3,300 mph.[1]

The fastest muzzle velocity ever recorded from any gun came from a much simpler weapon that used simple momentum conservation technology. A small object was measured to have a muzzle velocity of nearly 36,000 mph just after being fired by Sandia National Research Laboratories' 6 mm Hypervelocity Launcher. That's Mach 47—fast enough to go around the world in about two-thirds of a minute.

To obtain Mach 47 speed, the Hypervelocity Launcher uses three separate acceleration stages. In each stage, a big fast-moving object strikes a smaller one, accelerating the smaller object by the ratio of each item's respective mass. First a chemical explosion sends a heavy piston down a hydrogen-filled cylinder. As the piston travels it compresses the hydrogen until the cylinder pressure becomes so great that it ruptures a seal on the far end of the tube. The hydrogen then flows into a second, narrower tube that causes the gas to flow even faster. The gas accelerates a second piston, which moves forward at a brisk 16,000-mph pace until it slams into a relatively small quarter-inch-thick metal disk. The disk picks up the full momentum from the heavy, fast-moving piston and shoots out of the gun at nearly 36,000 mph.

The reason for building the Hypervelocity Launcher in the first place was to simulate the effect of space debris impacting spacecraft. To simulate the collision, a projectile is fired from the Hypervelocity Launcher and crashes into a stationary aluminum target that simulates a space vehicle. The impact triggers an ultra-high-speed camera that records the details of the collision for scientists to analyze. NASA hopes such simulations will help them design spacecraft that can withstand such impacts.[2]

THE TECHNOLOGY OF RAIL GUNS

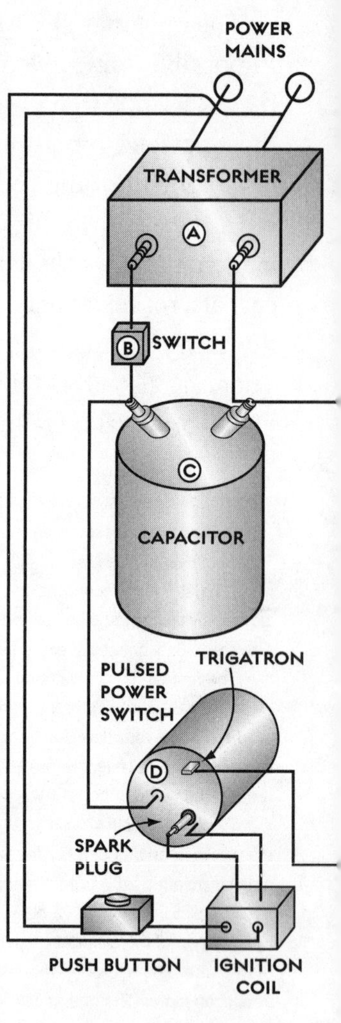

A rail gun is a non-combustion-powered cannon that uses electromagnetic effects to accelerate a magnetic shell to very, very high velocities toward a target. Basically, a rail gun uses the power from a rapidly discharging capacitor bank (not unlike the magneformer) to accelerate projectiles down two parallel conducting rails—hence the term "rail gun."

Homemade rail guns are another example of how clever amateur scientists exploit physicist Michael Faraday's nineteenth-century discoveries. Faraday calculated that a moving magnetic field causes electricity to flow in a conductor (and vice versa—electricity moving in a conductor induces a magnetic field around it). This is the reason motors, generators, and rail guns cause motion. Actually, all of these items are close to the same thing, just laid out a bit differently. The rail gun works by utilizing a closely related magnetic phenomenon termed the "Lorentz force" to shoot a projectile.

In any rail gun, there are two conducting rails connected to a power source, and a projectile is placed between the two rails, forming a conducting bridge. When the power is applied, current (typically, lots of it) will flow through the rails and the projectile.

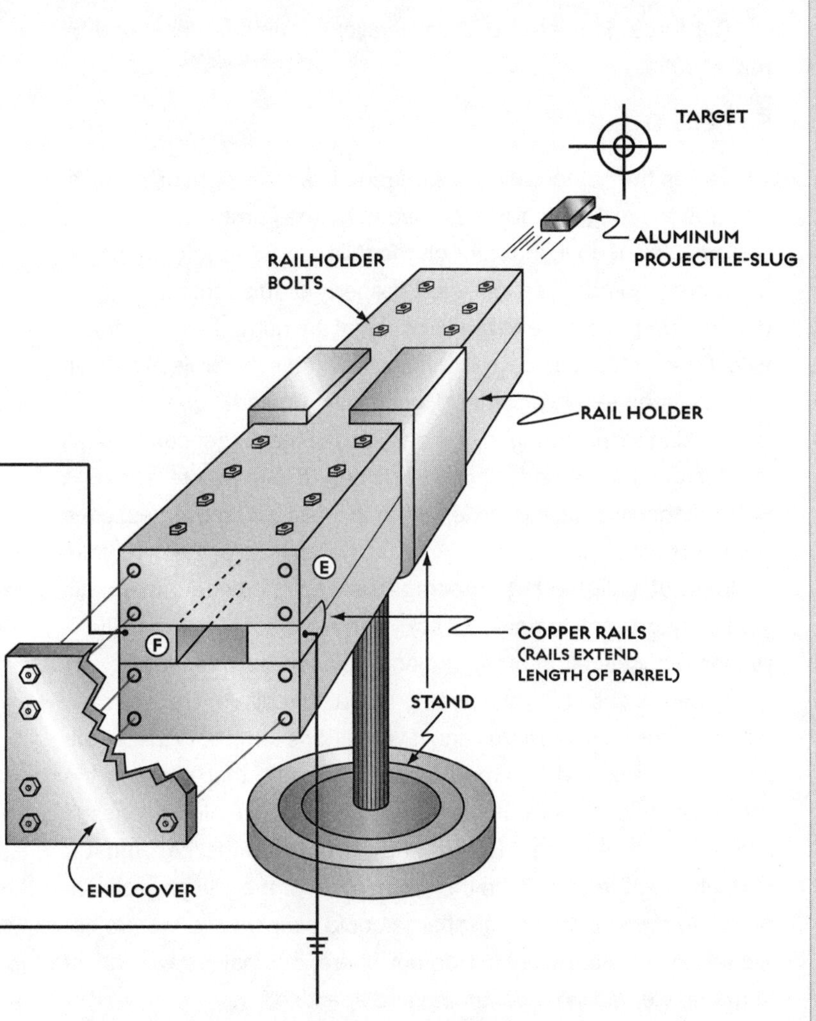

1. Transformer **(A)** changes capacitor bank **(C)**.

2. When fully charged, transformer is connected by opening switch **(B)**.

3. Firing push button is depressed and trigger current flows through high-volume pulsed power switch **(D)**.

4. Capacitor dumps energy to rails **(F)** inside gun barrel **(E)**.

5. Magnetic forces accelerate slug toward target.

The force generated by two parallel currents is described by this equation:

$$\mathbf{F} = \mathbf{i} \times \mathbf{L} \times \mathbf{B}$$

where F is the net force, i is the current, L is the length of the rails, and B is the magnetic force produced by the current.

The current coursing through the rails produces a force, called the Lorentz force. The significant property of this force is that it is directed perpendicular to the direction of the current and magnetic field. Given enough juice, this force accelerates the projectile down the rails at extreme velocities. Hence it's a powerful gun.

Looking at the equation, it is easy to see that larger currents will yield larger forces and higher velocities. If even more speed is desired, then increasing the length of the rails will also increase the duration of the applied force and thus the velocity of the projectile.

In a nutshell, the big amperage sets up a powerful magnetic field that acts on two parallel tracks and a projectile. This results in the rapid acceleration of the projectile placed between them.

It sounds easy to make a rail gun, but in reality there are many complexities involved in making it work. For example, in theory the capacitor bank is discharged through one rail, then through the projectile, and finally out the other rail. But in reality it doesn't always work like that: Before the Lorentz forces can muster enough effect to overcome the momentum of the idle projectile to start it in motion, the big discharge could simply weld the projectile to the rail. So it wouldn't go anywhere, but instead would just sit there, welded in place, arcing and sparking.

Therefore, the electromagnetic gun maker has to include a way to jump-start the projectile into motion just before the juice is applied. Often, a big push of air is the preferred way to start things off. An electrically operated solenoid valve is opened to allow a rush of compressed air or nitrogen from a storage tank to impinge upon the projectile. A split second later comes the electromotive force to accelerate the projectile to Mach speed. <

> > >

Sam Barros is a twenty-two-year-old engineering student originally from Brasília, Brazil. His current group of friends in the northern Michigan community he now resides in usually don't understand the stuff he builds or why he builds it. Most don't have a clue what he's doing, nor do they have the kind of electronics background to comprehend much about it even after he explains it. So Barros reaches out to others in the Technology Underground via the Internet.* Online, he's had a chance to talk with a lot of people who understand and respect what he does.

After his family left Brazil, Barros lived in the Netherlands, where he studied science of all types. From there, he moved to Michigan to study engineering. He is a man with a host of interests, and he is well known for his extreme technical hobbies. He is the man behind the website called PowerLabs.org, one of the most popular sites dealing with unusual and powerful home-built technologies.

PowerLabs contains detailed descriptions, photos, and videos of his many Technology Underground projects. A few of the interesting facts and descriptions found on PowerLabs involve his chemical synthesis of rapidly oxidizing chemicals (explosives and rocket fuels), Wimshurst machines, Tesla coils, high-power pulsed lasers, air guns, microwave guns, magneformers, and cryogenic experiments using liquid nitrogen. (The last category includes the spectacular but seemingly unwise liquid nitrogen baseball bat cannon.)

Making the cannon entailed cutting the end off an aluminum baseball bat. Liquid nitrogen was placed inside the bat and the

---

*For a great many citizens of the Technology Underground, much of their technical communication and human interaction with like-minded people takes place virtually, through forums and chat rooms on the Internet. For some of the most esoteric technologies, the chances of finding a lively community of enthusiasts with similar interests geographically nearby are nil. But the Internet compresses time and space, making communication easy, if not always accurate and honest.

open end sealed by tightly corking the end with a stopper. Barros held the bat shotgun style while aiming the stoppered end at a target. As the liquid nitrogen boiled inside the hollow bat, pressure built until the stopper ejects from the opening with sharp report and enough force to demolish a CD case.

With so many forays into the Underground, it's not surprising—in fact, it was probably predestined—that Barros would have at least one close call. His major mistake occurred in 1998 while performing perhaps the most dangerous of all extreme tinkering activities, rocket engine building.

"I was sixteen years old," Barros says. "I was interested in pyrotechnics, rockets, explosives, and so on. I started off with model rockets."

He moved past models quickly and started experimenting with home-built rocket motors. His first effort was very successful for an amateur with little experience or outside help. He fabricated a small hybrid rocket with an engine fueled by a mixture of butane, propane, and nitrous oxide. It was fast and powerful enough to generate several Mach diamonds[3] on its exhaust and was loud enough to cause the windows in his garage to shake while it was running on the test bench.*

"Then I started experimenting with solid-fuel rocket engines," he said. "First sugar and potassium nitrate, then I tried perchlorates, metal powders, nitrocellulose, and double-base powders. These small prototypes produced incredible thrust for their size."

He decided to push limits and proceeded toward more

---

* Mach diamonds are a radical tinkerer's best friend. Under the right circumstances they signify perhaps the most sought-after result of all—the attainment of supersonic speed. When a rocket or projectile goes supersonic, the area directly behind the rocket nozzle or projectile often exhibits a visible shock-wave pattern caused by the fast-moving exhaust "banging up" against the surrounding atmosphere. If a rocketeer or rail gunner's experiment goes well, the supersonic exhaust or slipstream pattern may resemble a series of diamond shapes, one right after another. This signals that supersonic speeds have been attained. The diamond-shaped patterns are termed "Mach diamonds" or, more technically, "N-shaped shock waves."

advanced designs and assemblies in quest of his own high-performance rocket engine. After various combinations of fuel-oxidizer chemicals and engine designs failed to produce the level of performance he was after, he decided a high-explosive-based propellant was the answer. Usually these engines employ what chemists term a "double base" mixture, in this case a combination of nitrocellulose and nitroglycerine, along with the usual fuel and oxidizer. Barros didn't have a method of acquiring the nitric acid he needed to make either of the nitrogen-based explosives, but he did know how to make a somewhat different type of high explosive.

"I made it, tested it, and thought it was safe," Barros says. "[But then] I was working with another batch that happened to be far more sensitive than the one I tested originally. I tapped the glass container it was inside to get the rest of the powder to come out and about three grams detonated on my hand.

"It ripped my thumb and index finger wide open," he recalls. "The explosion peeled the skin from most of the palm of my hand and peppered my arms, eyes, and face with pulverized glass fragments.

"An ambulance took me to the hospital. It was the most embarrassed I have ever felt in my entire life, and it cost a lot, too. None of the glass fragments punctured my eyes, though. Today, all I have are scars on my right index finger to remind me of the experience, but it could have been much, much worse."

He says he learned his lesson on that one. "I still perform the experiments I think are worthwhile, but I won't risk my safety like that again."

Among PowerLabs' most popular projects are the Linear Electromagnetic Accelerators, otherwise known as Rail Guns Versions 1.0 and 2.0. The rail gun project began when a major manufacturer of electric components heard about Sam's interest in and adeptness with electrical discharge experiments. Capacitors are required in magnetic pulse and motion experiments because those projects need a way to produce intense magnetic fields to induce movement or cause deformation. Barros tried out a whole menu of projects involving strong magnetic fields, including a solid-state can crusher, a disk shooter, a single-stage Gauss rifle, and, most significant, a multistage linear magnetic accelerator.

A sales representative for a capacitor manufacturer provided information regarding the results of some tests at his company in which currents as high as 40 kiloamperes were obtained from a single electrolytic capacitor. A bank of electrolytics would provide enough energy to give a device such as a rail gun a real shot at reasonable performance. But a capacitor stockpile consisting of such large components would be well outside his budget. Amazingly for Barros, the company not only helped him design a capacitor bank that would meet his theoretical rail gun needs—3.2 kV, 16,000 J, and the capability to deliver a current pulse of around 100 kA—but also gave him a shipment of large electrolytic capacitors for free. With those capacitors, Barros constructed the PowerLabs Rail Gun version 1.0.

The gun's capacitor bank was very large for an amateur science construction—thirty-two Cornell-Dubilier inverter-grade capacitors, each rated at 6,300 μF at 400 volts. Each capacitor was the size of an oversized soda can, and each can, when charged to 400 volts, held 500 joules of energy. Altogether, the big capacitor collection stored 16 kilojoules of energy. When the cap bank discharged, it was pretty intense. Says Barros, "It sounded like a handgun when it went off. The higher-powered shots had the crack you hear when a projectile breaks the speed of sound. The mechanism itself wasn't very loud. Most of the sound came from the projectile traveling through the air."

The rail gun has been enlarged and upgraded several times since, and it's clear that version 2.0 has potential for a lot of muscle. If any sizeable fraction of the 20,000 joules stored in the capacitors gets transferred to the rail gun's projectile as kinetic energy, it could take down a bank vault door.

> > >

It's past 1:30 A.M. again and the Internet rail gun forum postings are entertaining tonight:

After reading some posts on here, I was wondering: Who all has made a rail gun that works?

>> Hi. I've made a rail gun that "works." It was powered by 4 photoflash caps and could send a 1" square piece of aluminum foil folded to ⅓ inch by 1" about 5 feet across a table. Maybe someday I will make a better one. . . .

>> In high school I made a small one with 2.5 kJ of electrolytic capacitors. Copper rails, .25" spacing, .125" high. I switched it on with a big vacuum relay, and shot a carefully made projectile through a recycling bin. . . .

>> I have completed a larger railgun. It is powered by a 4,700 μF capacitor at 800 V and is switched on by the projectile making contact with the rails. . . .

When you first started logging in to the high-voltage hobbyist forums, the discourse was usually fairly civil. Through the postings of other experimenters, it's where you gathered the bits and pieces of technological arcana that permitted you to build your first coil gun. It's where you could buy decent capacitors without breaking your budget. Over time, though, the forum dialogue has become more cantankerous. The quality of communications and believability of your correspondents has begun to seem more and more questionable. Virtual fights have started to break out in some of the chat rooms, and a few of the forums are littered with traded insults and erroneous information. Evidently, there are some big egos involved. Tonight's postings seem a little more strident than normal.

[He] is a DIY hobbyist and a jack of all trades. He dabbles a little in everything but doesn't know a whole lot about anything. His website boasts the contents as science. It is obvious that most of his newest projects are not

> tailored for the scientifically inclined, but for attracting newbies and the laymen with dazzling "wow!" pictures and videos. He thinks he "knows it all."

It is interesting, sometimes even fun, to read the verbal jousting taking place between those separated by thousands of miles and lurking safely behind the anonymity of a computer screen. But in the technical forums, all you've got is your reputation, and people fight like dogs to maintain them.

10. ROBOTS

After half a year of intense thought, manual labor, and monetary investment, you've reached the moment of truth. You are standing on the raised driver's box, an elevated platform made from wooden planks supported by aluminum scaffolding. From here, you see down into every corner of the steel and Plexiglas arena where Webber, your warrior robot—the product of your blood, sweat, and tears—is about to enter mortal combat with another robot.

From the driver's platform, you look down upon Spin Doctor, the bot you will be fighting in a matter of moments. As you size up Spin Doctor in action, you slowly come to the unhappy realization that, frankly, your opponent's robot looks a whole hell of a lot meaner and stronger than yours.

It appears to be an aluminum cylinder standing on end, about the size of a quarter-keg of beer. Two or three furnace-forged, cobalt-alloyed, tool-grade steel hooks jut out of the body. A set of tiny motorized wheels whir around underneath the steel skirt formed by the cylinder. Motionless, it doesn't look like it can do too much.

When energized, however, its electric motor, originally used as a starter motor on a Volvo truck, applies a powerful circular torque to the heavy metal barrel. The Spin Doctor's body starts to twirl, faster and faster. The barrel spins so fast that it becomes simply a blur, and the steel hooks look like a solid disc, a solid plane of destructive, cutting energy. But it's the noise that really scares you, something like a tornado.

You force yourself to calm down and think about it. You know what's under Webber's hood; you know exactly what your bot can do. Webber is built to be a gamer—aggressive, destructive, and dangerous. The recently healed-over scars on your arms and shins—small nuisance wounds inevitably sustained in the course of building a powerful metal fighter—attest to that.

You spent the better part of several weeks working things out on paper—making notes, recording ideas, sketching out possibilities. Your notes begot scribblings, which begot drawings, which begot a detailed half-scale cardboard model. Then you spent the majority of your free time over the last several months in the garage, working on the real thing. The drawings became parts made from steel, aluminum, and rubber. The parts became bigger assemblies, the assemblies were bolted and wired together, and now here is your bot—a "saw-bot"—powered up and ready to go.

The house lights go down, the arena lights come up; it's time to rumble. At the referee's signal, you jerk the joystick on your remote control full forward, switch on the carbide-tipped fireman's rescue saw jutting out from Webber's rear like a wasp's stinger, and the battle is joined.

After the first closing, your saw leaves a gouge in Spin Doctor's armored side, while its whirling hooks crash into your tires. With this much horsepower, metal-cutting potential, and unrestrained momentum on both sides, it's clear this battle won't go the full three minutes.

You back Webber up a few feet and then give full power to the amperage controller. The batteries on board the robot shunt as much amperage as they can to the DC drive motors. With all the momentum the steel-bodied robot can muster, it crashes into Spin Doctor full force, leading with its armored plow. The Doctor is knocked off balance, wobbles, and finally topples over on its side. It's time to go in for the kill.

> > >

This is how it all comes together in the Underground world of warrior robot fighting. To the practitioners, it smacks of the 1999 movie *Fight Club,* where seemingly normal, mild, sensible men such as Edward Norton answer the primal call to get in touch with their masculinity through clandestine bare-knuckle fights. Robot fighting is a world of simple rules and brutish conflict. It's a way to prove manliness. But, unlike a fistfight, this offers intellectual stimulation, substitutes replaceable mechanical parts for blood, and avoids the whole broken-bone thing. Still, it's violent, aggressive, and a little nuts. It's part NASA, part NASDAQ, and part NASCAR.

In this world, like in NASCAR auto racing, just about everybody is part of a team. It takes a combination of people with different skills to build a top-notch fighting robot. Some teams have nihilistic-sounding names such as Death by Monkeys, Foaming Rampage, and Team Toad. Other team names simply sound threatening—Team Whup-Ass, The Inferno Lab, Team Killerbotics. Some go the dripping-with-irony route—Team Huggy Bear and Team Happy Robot. For all the teams, it's technical, greasy, and ambitious work.

Building a fighting robot requires the development of a vast amount of rather arcane knowledge, from how to build a proportional motor controller to how to choose the right aluminum alloy for a particular task. After a while, each team member personally understands every aspect of his mechanical progeny, from its DC permanent-magnet motors to the FM radio transmitter-and-receiver system that controls it.

Although they appear very complicated, robots are actually fairly straightforward in terms of architecture. How hard they are to build depends on the type and quality of machine tools and welding equipment used by the machine's makers. And how expensive they are depends on the builder's ability to scrounge and tinker.

THE TECHNOLOGY OF
FIGHTING ROBOTS

The basic, overarching purpose when building a warrior robot is to design it with the ability to do substantial damage to another robot. Certainly, this is a broad subject, and it provides a rich, fertile ground for exercising the most mischievous facets of the builder's imagination. Of course, unlike, say, a cowboy knife fight, there are rules—a typical warrior robot tournament is tightly regulated.

The robots are usually categorized according to the type of weapon they employ to wreak destruction. Some robots disable their opponents by cutting through armor to make minced metal out of the critical drive and electrical gear inside. Others disgorge huge amounts of energy to rotating flywheels and then bang the energy-laden spinner into opponents in an effort to shock-load the opponent into submission. Still others make use of electrical or pneumatically induced linear motion to flip their competitor head over heels and knock it out in the process. There is a taxonomy of sorts that can be used to describe the more common types of robotic weaponry:

Cutting blade weapon: Fireman's rescue saw **(A)**

Electric motor to turn saw **(B)**

Batteries to power weapon motor **(C)**

Drive train: Motors, power transmission, wheels **(D, E, F)**

Batteries for drive motors **(G)**

Radio control components **(H, I)**

Drive motor **(J)**

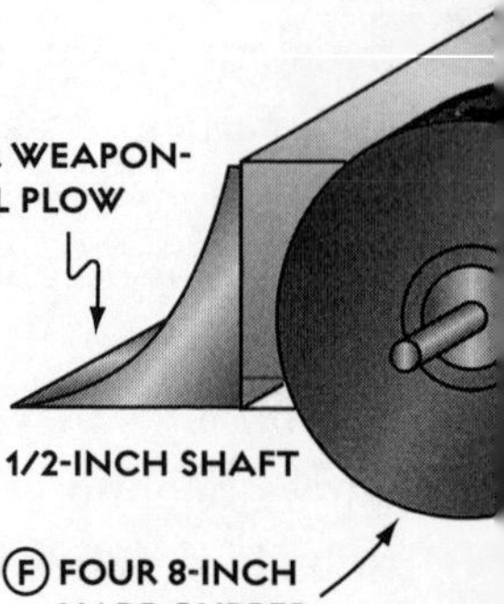

BMWS. *BMW* is an acronym for "battery, motor, and wheels." That's all that's there, just the raw power of mighty engines with grunting torque encased in a hardy armored shell. The simplest and most straightforward of all fighting robots, BMWs use the momentum developed by their own mass and speed to simply smash into their opponent. They may have rams or wedges in the front, or possibly steel spikes, but that's about it.

In the first *Godfather* movie, the Corleone family had two "soldiers" to do their dirty work, torpedoes named Brasi and Clemenza. If a BMW was a person, he'd probably look just like a Corleone thug: very large, mean, and violent, without a trace of finesse. A BMW has no spinning disks to balance and adjust, no tungsten-carbide cutting wheels that might throw a drive chain. The builder of a BMW simply puts all his effort and money into big motors, big wheels, and big batteries. The robot develops as much torque and speed as it possibly can, given its weight-class restrictions, and then muscles opposing robots into

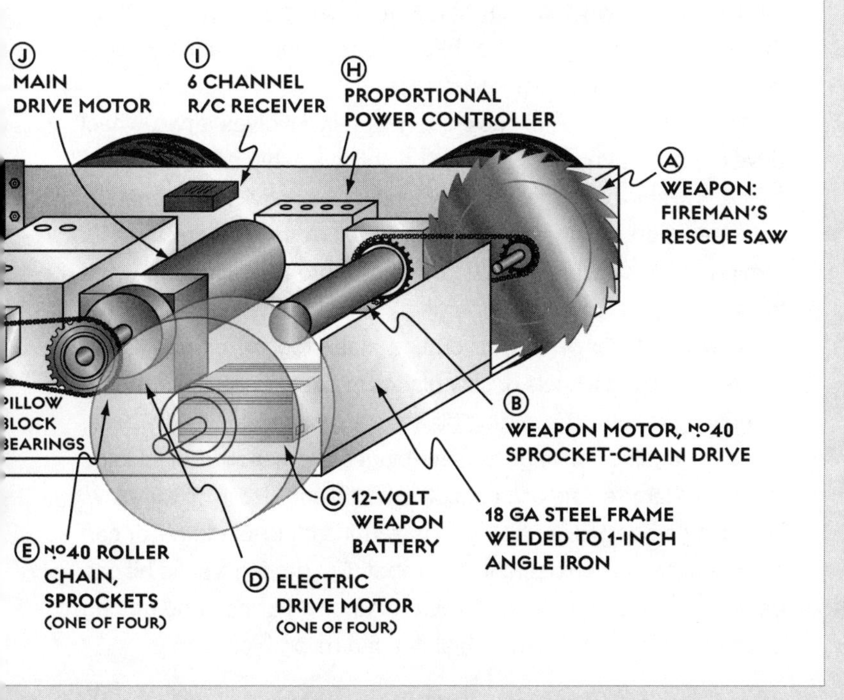

the arena walls and corners. This can be very effective, but success depends on driving skills and the ability to absorb punishment as well as dish it out.

The main strategy for a BMW is to transfer its kinetic energy (that is, the energy it possesses by virtue of its speed and mass) to its opponent in a way that is relatively painless for the BMW but really destructive to the other robot. This is best accomplished by hitting vulnerable parts with a ram or plow attached to the front of the BMW.

SPINNERS. A spinner robot uses the mass and velocity of a rapidly spinning outer shell to impart fantastic amounts of angular momentum and kinetic energy to opponents. The heavy shell is what engineers call a flywheel, a device that stores energy as a consequence of its mass and rotational velocity, that is, its angular momentum. The larger the mass, the more energy the spinning device contains. The faster it spins, the more energy the device contains. The idea behind a spinning robot is to attack with a fast-turning, heavy flywheel that has a hard, sharp protrusion. When the heavy spinning metal shell hits the opponent, the energy transfer is enormous and destructive.

THWACK-BOTS. A thwack-bot generally involves a two-wheel drive design that provides it with the ability to whip around in a circle. A thwack-bot turns quickly and powerfully. It is generally equipped with a long, stiff, and heavy tail. A thwack-bot uses its turning ability to whip its tail around and produce damage when the spinning tail contacts its target.

SAW-BOTS. Saw-bots, or cutting-blade robots, employ large, hardened cutting wheels or saw blades to cut through the armor of competing robots. The cutting blades are generally driven by belts, gears, or chains. Driving the saw blades are gasoline or electric motors, as large as the competition's weight and size rules will allow.

A single good, solid hit by a powerful cutting-blade robot can incapacitate or even decapitate a robot. But getting a solid hit can be difficult. The saw blades tend to bounce off armor, and getting good purchase to start the initial cut can be problematic.

LIFTERS. Robots with arms or forks designed to lift and throw opponents are among the most exciting and successful. A lifter has a mechanical arm, powered by either electrical motors or high-pressure compressed-gas cylinders, designed to get underneath the opposing robot. Then, in the blink of an eye, it lifts the opposing robot quickly, sending it careening across the arena floor. Few robots are built solidly enough to withstand being flipped more than a time or two.

HYBRIDS. There is nothing in the rules that precludes combining more than one mode of attack in a single machine—for instance, placing a lifting fork in front and a metal saw in the back. The problem with this approach is the weight limits in place at all tournaments. Most builders feel it is better to concentrate on a single, powerful offensive strategy than to divide their time, money, and weight allowance between multiple ideas. <

BOTBASH

Fighting robot tournaments take place in a very wide variety of venues—sometimes in rented arenas, sometimes in an old warehouse, sometimes in a big hole recently bulldozed just for this purpose on abandoned railway property, with spectators looking down from the earth berms mounded around the hole.

A tournament called BotBash was typical. It happened in a beat-up old warehouse in the industrial outskirts of Phoenix. The event's promoters set up a large cage, made of expanded metal and reinforced with angle irons and rebar where necessary. This was the pit where the robots attempted to clang and bash each other with as much momentum and energy as they could muster. When one robot became immobilized or perhaps reduced to nuts, bolts, and steel splinters, that match was over and the next one began. Over the two days of the event, there were umpteen bouts where robots such as Ziggo, Toe-Crusher, and Wedge of Doom fought one another to mechanical stasis, a sort of intellectual and semi-civilized dogfight with iron pit bulls.

A crowd of 150 or so metal-bending robot builders surrounded the fighting pit. Who were these builders, and where do they come from? They were engineer and technical types, almost exclusively male, with an edginess of sharp metal about them. A lot of them wore T-shirts with the names of their teams and pictures of their robots on them. During the combat, they pressed up against the robot-fighting area and surrounded the pit on all sides, completely absorbed in the metal-against-metal carnage within.

The excitement down near the fence was intense. After a really, really good match, one in which the robots and their remote drivers gave it all they had in terms of energy and ability, the other builders showed their admiration. They banged loudly and rhythmically on the metal fence encasing the pit, in collective praise.

Next to the fighting pit were the work areas—long tables lit by portable halogen lamps whose harsh glare filled the warehouse with giant moving shadows. The builders hunched over their creations, borrowing hammers, arc welders, and soldering guns from one another in order to put their freshly wrecked robots back into fighting shape before the next match. The builders are resourceful and can repair almost any mechanical affliction.

Yet not all of the damage sustained by the robots in battle is repairable. When a gas-powered hammer comes down in just the right (or wrong, depending on whose ox is being gored) way on a control circuit, or an ultra-high-velocity revolving chain does too much damage to a main axle, there may be no way to fix the robot. So, in the course of the tournament, the back walls gradually accumulate piles of junk—here a bunch of twisted wheel rims, there a mound of smashed bearings that will never come close to turning freely again, everywhere scraps of burned-out electronics parts and inoperative pneumatic cylinders—all tossed into heaps for the cleanup crew to cart away. What took a builder days, perhaps weeks, to drill, shape, and weld into a precision part was reduced, in one painful moment, to junk, useful for nothing but sandbag filler or as a boat anchor.

In the back of the BotBash pit area worked one of the most successful builders, Jason Dante Bardis, still boyish-looking in his mid-twenties, dark-haired, and personable. Like a lot of the other builders there, Bardis started building robots as a youngster—first from plastic LEGO toys, then from scraps of metal and wood, and now from aircraft aluminum alloys and space-age carbon-fiber composite materials. Jason has a doctorate in mechanical engineering from a top school and he's a whiz with motors, machine tools, and materials, all of which makes him a pretty crafty guy when it comes to building robots. He built his own troop of fighting machines, more than half a dozen in all. Bardis's robots range from a several-hundred-pound rolling robot outfitted with hammers to his lightweight champion, Dr. Inferno Junior.

Jason calls his robotic menagerie Dante's Infernolab. His best robot, in terms of fame and success, is Dr. Inferno Junior. The

Doctor was the overall winner in the lightweight division on television's *Battlebots,* a nationally televised program that was the robotic equivalent to the Super Bowl when it was on the air.* To get that far, Dr. Inferno Junior had to win bout after bout after bout against the very best fighting robots and robot builders in the world.

At BotBash, Bardis was hard at work, focused intensely on his task despite the acrid fumes of welding flux and burning electrical insulation that overwhelmed the pit. Oblivious to distractions, he finished making the last critical adjustments on his new ultra-lightweight robotic creation, Hell on Wheels. HOW is a two-wheeled bot with a couple of ominous-looking prongs sticking out from the front. It is a fast, smoothly operating bot with a fair amount of armor. HOW is basically a hardened box with wheels attached to motors attached to batteries. But what this robot does have is quite a load of power—its power-to-weight ratio is excellent.

---

*In the late 1990s, one of the highest-rated shows on cable television was called *Battlebots.* The program consisted of a televised three-minute robotic cockfight between two radio-controlled vehicles.

It was interesting to watch at first and commanded a pretty impressive portfolio of viewer demographics. After a while, though, much of the general public grew tired of watching one metal robot after another wheel around a Plexiglas-enclosed arena dodging, weaving, and ultimately breaking. It became monotonous, one bot simply clanking against another, despite attempts at creative editing and sound effects. After the fourth season, ratings were such that network executives relegated *Battlebots* to Saturday night, which in television programming circles is where shows go to die. Then, with little fanfare, the plug was pulled.

In the wake of the show's cancellation, robot fighting contests and tournaments became local and regional affairs. They were gritty and low-budget for sure, but the wide dispersion and higher frequency of events meant that more people could try their hand. The rules for building bots were loosened and the activity, set free from the de facto television show standards, expanded from a small cadre of die-hard enthusiasts into a larger if not quite so serious community of robot builders. Since *Battlebots* left the airwaves, scores—maybe even hundreds—of low-overhead robot tournaments have come, stuck around a while, and gone. The best were sophisticated, professional affairs with hundreds of spectators on bleachers surrounding big clear cages made of Lexan polycarbonate. Some have been far more modest, consisting only of the builders themselves watching their machines duke it out in a crude pit dug into the ground. It's not fancy, but it is genuine.

Jason's twin DC drive motors are strong and unique. Jason devoted his design time and abilities to making a simple, rugged robot that rams into things with tremendous force and determination.

The Bardis-led team incorporates a good deal of high technology in its designs. The motors employed in most bots use electric currents to set up attracting and repelling magnetic fields inside the motor housing. These fields push the rotating parts of the motor around in an endless circle, producing torque and, therefore, motion. The rotating part of a motor is called the rotor, and the stationary part is called the stator.

Typically, robots use direct-current motors connected to battery packs for locomotion. In order to get continuous motion, a mechanism within the motor housing, called a brush, is used to sequence the application of electric current from the stator to the rotor so that the motor spins smoothly and in one direction. Brushes have been used in every DC motor since Edison.

But there are no brushes in Jason's robot motors. He uses the most advanced technology around—solid-state, computerized, brushless motors—in his armada of robots. Substituting solid-state brushless technology for the make-and-break action of regular motor brushes reduces the number of mechanical parts and hence the number of friction and wear points in the machine. As a consequence, Infernolab robots are tough, fast, and light. They hit hard and win often because they are able to put out enormous amounts of torque and power when needed. For its size, Hell on Wheels pushes harder than a copulating elephant.

The Infernolab members sat in the pit area working delicately with soldering irons under a portable spotlight. They were taking great pains to securely attach the battery cables to the terminal wiring strip that serves as the main cross-connect point for all the electrical connections on Hell on Wheels. It was somewhat tedious work, but vitally important, because all it would take is one loose wire to transform the robot from a metal-crunching, rip-snorting hellion to a mailbox.

Soon after, the elimination rounds started. The best bots became apparent quickly. Unfortunately for Bardis, Hell on Wheels had an unexpectedly early exit when it fell through a hole in the arena floor and smashed on the concrete below. But the lessons learned in battle served Bardis well in later events. His lightweight, aluminum-skirted alter ego, Dr. Inferno Junior, became one of the most successful fighting robots in history. It took the Battlebots Season 3 championship. And Bardis accepted an engineering position, designing and engineering electrical components, with the company from which he originally procured his robot motors.

On this night, Ziggo, Wedge of Doom, Herr Gepounden, Backlash, and Gamma Raptor looked like the ones to beat. Gamma Raptor is the brainchild of the Pitzer brothers, Bob and Chuck, of Chandler, Arizona. The Pitzers are engineers in their twenties who caught the fighting-robot bug early. They are so enthusiastic about the activity that they organize their own tournaments, including this one. Their bots have a reputation for toughness. Gamma Raptor is low and mean and brutish in appearance, and it looks just the way a fighting robot should look: solid. It's a titanium-sheathed robot with a pair of tank tracks on each side for locomotion, an array of metal arms designed to stab and gouge, and a fork to slide under the body of opponent robots. In fact, it faintly resembled a low-slung fork lift both in form and function.

Gamma Raptor is a type of fighting robot known as a lifter. This type of robot's mode of attack is to slide the fork, powered by a high-pressure gas cylinder, under its opponent and then lift. On board Gamma Raptor, underneath the titanium and aircraft-alloy aluminum armor, there is a reservoir of high-pressure gas contained in reinforced, fiber-wrapped plastic bottles. At the touch of a button on the driver's remote control console, a valve opens and the gas is released. Deep inside the robot, the expanding gas acts on a piston in a pneumatic cylinder that quickly and forcefully pushes a connecting rod outward, and in doing so, powerfully jerks the end of the pivoted fork upward. The power released by the gas is immense, and therefore the pivoting fork has the strength and velocity to flip an impressively heavy weight high in the air as well as end over end. Opposing bot drivers know this strategy and so make every effort to avoid the deadly fork, but Gamma Raptor is fast and maneuverable and often gets the fork home before there's anything that can be done. During the two days of competition within the warehouse, Gamma Raptor seemed just about unstoppable.

In a parallel bracket, another robot was making mincemeat out of the other metal mowers. Backlash is a "sawbot"—basically a movable platform that directs the motion of an incredibly large, case-hardened, fast-spinning, vertically mounted saw blade. Backlash is the brainchild of a veteran builder named Jim Smentowski who was formerly a Hollywood special-effects designer for George Lucas's Industrial Light and Magic.

In 1996, some friends from ILM told Smentowski about Robot Wars, and he went to see the carnage firsthand.* After seeing fighting robots with names such as La Machine and Biohazard, he was hooked.

"I did what only came naturally for about 98 percent of the guys who attended Robot Wars," he says. "I started thinking to myself, 'Hey, I could do that!' "

---

*In the early 1990s, a San Francisco–area performance artist and special-effects producer named Marc Thorpe started experimenting with the idea for a eight-wheeled remote-controlled vehicle for vacuuming and cleaning office buildings. It was a bust as a robotic maid, but it was very interesting to watch. So interesting, in fact, that its inventor trashed the vacuums and instead started attaching weapons of various kinds to the chassis.

Thorpe's idea was to build a mechanical warrior, or rather several of them, and allow them to fight in front of a paying crowd. In 1994, the first Robot Wars event took place in a converted aircraft hanger on an island in the middle of San Francisco Bay. It was a great success and inspired a number of followers such as *Battlebots* and Mechwars.

He designed various robots, each of which was better than the one before, until he came up with his best effort, the bot he calls Backlash. The saw blade on Backlash is made from an approximately 2-foot-diameter machined disk of aluminum mounted to a large electric motor. The motor is powered by a densely packed array of nickel-cadmium batteries. There are a couple of gnarly alloy-steel teeth that jut out from the aluminum disk, devouring the competition.

At the end of the day, Gamma Raptor and Backlash had battled through all the other competitors to reach the championship match, Backlash's saw versus Gamma Raptor's lifter. Each robot driver attempted a flanking strategy, trying to avoid the other's main weapons. The two jousted and parried, each trying to bring its front-mounted weapon to bear against the unprotected flank.

Both competitors hung tough. Neither one did much substantial damage to the other early on. But in this activity, things can turn quickly. Sensing a slight delay in orienting Backlash's saw, Gamma Raptor swooped forward and by the thinnest margin managed to get its lifter's fork tines under Backlash and lift it off the ground. The robots wrestled around awkwardly, and with Backlash off balance, the Raptor poked its metal protrusion, looking like a shark's fin, into a rip in Backlash's underside. With its opponent stuck on top, Raptor smelled blood and whirled around, carrying the trapped Backlash, its saw spinning impotently, and slamming it into the arena side wall several times.

Gamma Raptor has large, powerful, high-torque motors, and the Pitzers decided that it was time to use every last drop of power in those batteries. It was now or never. The electronics controlling Gamma Raptor flooded the motors with electric current. The driver sent full forward power to the tracked wheels on one side and full reverse power to the wheels on the other, sending the bot into a whirl. This spinning maneuver threw the trapped Backlash several feet into the air, and it came crashing to the arena floor.

Sensing the end, the audience hooted and pounded on the arena walls. But Backlash still had fight left. It spun up its giant metal cutting wheel again and charged headlong into the front of Gamma Raptor, right at its lifter. The rotating cutter caught a bit of Gamma Raptor's side and tossed it upside down, mangling the forks. Both robots had become weakened, with power in the batteries ebbing away, but they still fought on, albeit more slowly.

At the final buzzer, both bots were still standing. The match was a classic: a testament to their rugged materials, their solid welding and wiring, their competitive nature. Somebody had to win. In the end, the judges awarded Gamma Raptor the match and the championship on points. It could have gone either way.

After the tournament, the builders broke it down, both literally and figuratively.

"You know what I need to win?" a builder in the pit area asked. "It's not just motors and weapons. I need bag fries."

People looked confused, so he continued. "You know when you go to Burger King or McDonald's and get a combination at the drive-through? Well, when you finish off the fries and take the last bite of burger, you still wish there was a little more left, just a few more morsels, one last bite. Then you look in the bottom of the bag and there they are—the three or four french fries that fell out of the cardboard container and are sitting there in the bag, hiding from you. It's just what you need—it's that little bit extra that puts you over."

The crowd got it, and he nodded. "The robot that wins has a little something extra. The equivalent of a last-minute booster rocket, the afterburner. A little something extra. Bag fries."

conclusion

NOTES FROM THE TECHNOLOGY UNDERGROUND, CONCLUDED

From killer robots and supersonic rail guns to Nazi-scientist-developed pulsejets and Crusader-built catapults—once you get to know them, none of these machines is inherently frightening. It's what people do with them that is. The same krytron or trigatron that initiates an H-bomb explosion also forms and shapes metal products for industry. The same radiation dangerously spewing from a Z-pinch can also provide medical X-rays and oncology treatments. And the same linear magnetic acceleration that powers a rail gun can float a magnetically levitated train. The inhabitants of the Technology Underground recognize this and choose to do entertaining things with their technological prowess.

It is hard to deny the visceral satisfaction of observing the motion, and appreciating the momentum, impulse, and volume of a large catapult, a high-powered rocket, a warrior robot, or an air gun. It is impressive technology and fun to encounter. But beyond this straightforward, gut-level enjoyment, you might ask if this stuff has any real relevance. That is, does the Technology Underground—the machines, the participants, and the events—have any particular social consequence or impact?

First, consider the status of scientists and engineers in today's world. While they may be well paid, and possibly enjoy admiration within their own peer group, the fact is there are not many living scientists or engineers who have attained much in the way of mass-culture popularity. In fact, there are none.

Perhaps, on the odd occasion, an autograph seeker stalks MIT's Old Main in hopes of obtaining Marvin Minsky's or Steven Pinker's signature. But really, very few scientists need bodyguards to keep away the star-struck rabble. *The Tonight Show* has rarely, if ever, invited a famous engineer or scientist to sit on the couch and schmooze with Jay Leno.

Each year, a company called Marketing Evaluations polls roughly fifteen hundred Americans living in the continental United States to produce Q-scores, popularity evaluations of a potential spokesperson or endorser. Results from a recent survey, in order of their finish, include Bill Cosby (his 71 is the highest Q-score ever recorded), Albert Einstein (56), Mickey Mouse (44), Elvis Presley (33), Batman (13), LL Cool J (12), Donald Trump (11), George Steinbrenner (10), Howard Stern (10), the Teletubbies (9), Carmen Electra (9), Larry King (7), and Count Chocula (7). No living technology-related individual, with the exception of Bill Gates, pulls a Q score higher than Count Chocula.

It's not hard to name scientists from the first half of the twentieth century: Edison, Tesla, the Wright brothers, Einstein, Teller, Marconi, Carver, Carrier, Fleming . . . the list goes on and on. But sometime around 1950, science and invention lost its individualism. Progress became a product of large, corporate, methodical, and highly specialized work teams. Invention became the province of almost exclusively faceless and nameless groups and departments. Almost anyone can tell you who invented the telephone, radio, or lightbulb, but very few can say anything factual about the inventor of the television, the personal computer, or the Internet.

The Underground is a place where recognition and admiration for a job well done are freely given. There, machines and devices speak volumes about what is good and interesting about technology. And the more interesting and novel the machine is, the more glory accrues to its creator. Life in the Underground is merit-based and logical. It offers inspiration. People keep coming back for more, and it keeps them motivated to do more, improving upon improvements. It is an incubator, for ideas as much as for machines, and from it ultimately come some of the best scientists in the world.

Interesting machines leave a lasting impression, on individuals and on history. In some cases they rival famous people. Just as there were many famous scientists and engineers in the first half of the twentieth century, and few to none in the second half, so it is for the machines themselves. The famous-machine phenomenon used to be far more widespread and common than it is currently—at one time, machines could be as well-known as kings and presidents, better known than singers and actors. But for some reason, great machines no longer enjoy the mainstream popularity they once did. There are a few modern superstar machines, but not many.

Take the Rocket, for example. For the greater part of the period since the beginning of the Industrial Revolution, the single most famous machine in the world was an English steam locomotive known far and wide as the Rocket. It's been somewhat forgotten in recent years, but for nearly two centuries, the Rocket was as well-known as any machine has ever been—as famous in its time as Elvis Presley was in his. It became so by winning a contest called the Rainhill Trials way back in the fall of 1829.

It was a time of massive industrialization throughout Europe. When the first iron track between Liverpool and Manchester was just about complete, the company owners needed to choose a railroad engine to pull the cars. In a rare move that combined business insight and public relations acumen, they decided to hold a competition to award the contract for the type of engine to be used. This competition was called the Rainhill Trials and was pretty much arranged as an all-comers event for anybody and everybody who had a locomotive.

Public interest in Rainhill was tremendous—in fact, some historians compare it to the first manned trip to the moon, or the race for the South Pole. Each day of the trial, details were reported on the front page of newspapers around the world. Each day for several days, some twelve thousand to fifteen thousand people wandered out to the banks of the River Mersey to watch the Rocket take on the competition.

Scores of potential entrants had dropped out along the way, and the Liverpool and Manchester Railroad had but five true competitors to judge. But a stout and creative five they were: Cycloped, Novelty, Perseverance, Sans Pareil, and Rocket, the last of these built by George and Robert Stephenson.

Cycloped included a variety of technologies, most of them bad, first and foremost of which was a draft horse walking on an endless belt. Cycloped was described in the October 1829 issue of *Mechanics Magazine* as powered by "a squirrel-cage, supposing that the squirrel drives its circular cage round by treading on the outside instead of the inside." It left the competition after the horse fell through the floor onto the track.

Perseverance never really had a chance—the contest rules required that all engines reach a speed of 10 miles per hour, and no matter how hard Perseverance persevered, it could not make that speed, so it withdrew. Sans Pareil did not complete the trial either, as it cracked a cylinder early on.

The last engine to leave the competition was Novelty, the crowd's choice and the early favorite to win. It was faster and lighter than the other engines; in fact, it was a veritable bullet on rails, reaching an astounding 28 miles per hour. But it kept breaking down, at first fixably, but then completely and irreparably, and so it too was gone.

The Rocket was the only locomotive to complete the trials. It averaged 12 miles per hour, hauling 13 tons, and so chugged away with the £500 prize. The Stephensons were awarded the coveted contract to produce locomotives for the L&M, and the Rocket became the most famous machine in the world. People from Manchester to Massachusetts to Madras knew all about the Rocket, and in fine technical detail at that. This was the first little engine that could, and it did, becoming the first mechanical superstar of the Industrial Revolution.

Like the Rocket, the best radical technological inventions have their own names and personalities and are often discussed as entities by themselves, apart from their builders. For instance, the Subjugator, Old Glory, Dr. Inferno Junior, and Electrum are likely more famous than the people who created them.

So, if the machines of the Technology Underground are, like the Rocket, famous in their own way, does this make them important?

Emphatically, yes.

This may not be obvious at first. There are precious few important inventions that have passed directly from the Technology Underground into the fabric of modern life. For the most part, there are but a very small number of practical extensions, such as amateur flamethrowers, that have made their way into the daily lives of most people. But look deeper and you will see a significant impact. For the Technology Underground is a prep school of sorts, a place where a vast number of mainstream technologists and scientists have been inspired, encouraged, and launched.

High-power rocketry has influenced a large number of NASA scientists and served as a springboard to a far larger number of successful scientists and engineers in other fields. Any number of the young men and women in the warrior robot pit go on to great things in more traditional roles as engineers. The fabrication skills that a person learns welding and riveting a big air gun translate to leadership roles in factories, shipyards, and industrial plants. Much, perhaps *most,* of the progress and innovation that support and operate the technology of modern life comes from the thousands upon thousands of small companies and individuals making drive train components, hydraulic systems, and electrical assemblies.

These items don't come from government-funded high-tech labs such as those at Stanford, Sandia, Princeton, or Livermore.

For the most part, they don't come from General Electric, General Motors, or General Dynamics either. They come from individuals or from small, privately owned firms that do not have a single PhD or MBA on staff. Some of the people involved don't even have college degrees. The people in these organizations learned their craft and honed their skills less formally. They learned a lot of what they need to know hanging around Underground.

In the days of the itinerant electrifiers, a few paper puppets dancing to static electricity was interesting, and a demonstration resulting in a dead, smoking pigeon would be talked about for weeks. Of course, it's getting harder to impress people with electrical effects. The turning disks of a Wimshurst machine are old-fashioned and mechanical, and the idea of watching someone make three-inch-long sparks for an evening's entertainment is quaint and nostalgic. Still, the concept lives on, because technology continues to entertain.

notes

CHAPTER 2: THE TECHNOLOGY OF BURNING MAN

1. St. John, Graham. *FreeNRG: Notes from the Edge of the Dance Floor.* Melbourne: Common Ground, 2001.

CHAPTER 3: TESLA COILS

1. Bulman, Alan Davidson. *Models for Experiments in Physics.* New York: Ty Crowell, 1968.

CHAPTER 5: HURLING MACHINES

1. Paul, Jim. *Catapult: Harry and I Build a Siege Weapon.* New York: Villard, 1991.
2. Niel, Fernand. *Montségur: le site, son histoire.* Grenoble: Alliers, 1962, 222.
3. Gurstelle, William. *The Art of the Catapult: Build Greek Ballistae, Roman Onagers, English Trebuchets, and More Ancient Artillery.* Chicago: Chicago Review Press, 2004.

CHAPTER 6: AIR GUNS

1. Beeman, Robert. "Four Centuries of Air Guns" in *Air Gun Digest.* Northfield, Ill.: DBI Books, 1977.
2. Hutton, Charles. *Mathematical and Philosophical Dictionary* (London, 1795–96).

CHAPTER 7: FLAMETHROWERS

1. http://en.wikipedia.org/wiki/Flamethrower

CHAPTER 9: RAIL AND COIL GUNS

1. "Experiments with the Green Farm electric gun facility" McNab, I. R., LeVine, F., Aponte, M. In *IEEE Transactions on Magnetics,* January 1995, volume 31, issue: 1, part 1.
2. "Orbital speed in Albuquerque-space-debris collisions simulated in laboratory." *Discover*, July 1994.
3. Sutton, George P., and Oscar Biblarz. *Rocket Propulsion Elements,* 7th Edition. New York: John Wiley & Sons, 2000.

further reading

Robert Beeman. *Four Centuries of Air Guns.* Northfield, Ill.: DBI Books, 1977.

A. D. Bulman. *Models for Experiments in Physics.* New York: Crowell, 1968.

Margaret Cheney. *Tesla—Man Out of Time.* New York: Barnes and Noble, 1993.

William Gurstelle. *The Art of the Catapult.* Chicago: Chicago Review Press, 2001.

Charles Hutton. *Mathematical and Philosophical Dictionary.* London: Johnson and Robinson, 1795.

Jack Kelley. *Gunpowder.* New York: Basic Books, 2004.

R. Lindeburg. *Mechanical Engineering Reference Manual.* Belmont, Calif.: Professional Publications, 1995.

Terry McCreary. *Experimental Rocket Propellant.* Murray, Ky.: author, 2000.

Jim Paul. *Harry and I Build a Siege Weapon.* New York: Villard, 1991.

Gerhard Schaefer. *Gas Discharge Closing Switches.* New York: Plenum Press, 1991.

Brad Stone. *Gearheads: Turbulent Rise of Robotic Sports.* New York: Simon and Schuster, 2003.

George P. Sutton. *Rocket Propulsion Elements.* New York: Wiley, 2000.

Graham St. John. *FreeNRG: Notes from the Edge of the Dance Floor.* Melbourne: Common Ground, 2001.

acknowledgments

This book is the result of two years spent spelunking the Technology Underground. I was continually impressed by the creativity, originality, and technical intelligence of the people I met.

It takes a while, a long while, to get through the book writing process, and I thank everyone for their patience and continuing support. So, much thanks to those whose names appear in the book—they are without exception a group of interesting, brainy, and unique individuals. Also, given my limited powers of memory, to those people who helped me but whose names do not appear here, I thank you and apologize for the oversight.

Thanks to Jane Dystel, Elissa Altman, Jennifer DeFilippi, and Chris Pavone, giants all of the publishing industry, as well as the staff at Clarkson Potter who brought this book to reality. Thanks to friends, colleagues, individuals, and organizations who provided such valuable support and assistance: my friends, sons Ben and Andy, the Bakken Museum Staff, the World Championship Punkin Chunkin organization, LDRS-21 organizers, the Burning Man Media Mecca team, Dr. Terry McCreary, Harold Reed (the Penis Doctor), and Casimir Sienkiewicz. Finally, Karen Hansen helped, encouraged, and inspired me, and I owe her so much.

index